4th International Conference on Geological, Geographical, Aerospace and Earth Sciences (AeroEarth 2016)

IOP Conference Series: Earth and Environmental Science Volume 38

Bali, Indonesia
4 – 5 June 2016

Editor:

F. L. Gaol

ISBN: 978-1-5108-2814-8
ISSN: 1755-1307

Printed from e-media with permission by:

Curran Associates, Inc.
57 Morehouse Lane
Red Hook, NY 12571

Some format issues inherent in the e-media version may also appear in this print version.

Printed by Curran Associates, Inc. (2016)

For permission requests, please contact the Institute of Physics
at the address below.

Institute of Physics
Dirac House, Temple Back
Bristol BS1 6BE UK

Phone: 44 1 17 929 7481
Fax: 44 1 17 920 0979

techtracking@iop.org

Additional copies of this publication are available from:

Curran Associates, Inc.
57 Morehouse Lane
Red Hook, NY 12571 USA
Phone: 845-758-0400
Fax: 845-758-2633
Email: curran@proceedings.com
Web: www.proceedings.com

TABLE OF CONTENTS

Preface

The 4th International Conferences on Geological, Geographical, Aerospace and Earth Sciences 2016 (AeroEarth 2016), was held at Four Points by Sheraton Bali, Kuta, Bali, Indonesia during 04 – 05 June 2016.

The 1st AeoroEarth 2013, 2nd AeroEarth 2014 and 3rd AeroEarth 2015 were already held succefully in Jakarta, 2001; Kuta Bali, 2014 and Jakarta, 2015 respectively.

The 1st AeoroEarth 2013, 2nd AeroEarth 2014 and 3rd AeroEarth 2015 have already been published in EES IOP: http://iopscience.iop.org/1755-1315/19/1, http://iopscience.iop.org/1755-1315/23/1, http://iopscience.iop.org/1755-1315/30/1 respectively.

The theme of AeroEarth 2016 is "Recent Trends in the Geosciences"

Earth provides resources and the exact conditions to make life possible. However, with the advent of technology and industrialization, the Earth's resources are being pushed to the brink of depletion. Non-sustainable industrial practices are not only endangering the supply of the Earth's natural resources, but are also putting a burden on life itself by bringing about pollution and climate change.

A major role of earth science scholars is to examine the delicate balance between the Earth's resources and the growing demands of industrialization. Through research and development, earth scientists have the power to preserve the planet's different resource domains by providing expert opinion and information about the forces which make life possible on Earth.

We would like to express our sincere gratitude to all in the Technical Program Committee who have reviewed the papers and developed a very interesting Conference Program as well as the invited and plenary speakers.

This year, we received 68 papers and after rigorous review, nine papers were accepted.

The participants come from six countries. There are two Parallel Sessions and four Keynote Speakers.

It is an honour to present this volume *of IOP Conference Series: Earth and Environmental Science (EES)* and we deeply thank the authors for their enthusiastic and high-grade contribution.

Finally, we would like to thank the conference chairmen, the members of the steering committee, the organizing committee, the organizing secretariat and the financial support from the conference sponsors that allowed the success of AeroEarth 2016.

The Editor of the AeroEarth 2016 Proceedings.

Dr. Ford Lumban Gaol
Bina Nusantar University

IOP Conf. Series: Earth and Environmental Science **38** (2016) 011001 doi:10.1088/1755-1315/38/1/011001

The 4th
Services
The 4th
Geological,
Four Points

IOP Conf. Series: Earth and Environmental Science **38** (2016) 011001 doi:10.1088/1755-1315/38/1/011001

IOP Conf. Series: Earth and Environmental Science **38** (2016) 011001 doi:10.1088/1755-1315/38/1/011001

Keynote Speakers

1. Prof. Ee-Peng Lim

Title: Multimodal Sensemaking using Social Media Data

Abstract:

As social media becomes an integral part of daily lives, it captures interesting user-generated content and behaviour data that can be sensed and analysed. While social media companies use the insights learnt from such data to improve their user interface and experience, there are many other interesting insights that help us improve urban environment and public services. Social media data also offers a cheap and scalable approach to perform sensemaking on the urban environment. In this talk, we will showcase a few ongoing research projects in the Living Analytics Research Centre (LARC) which focus on multimodal sensemaking using social media data. The talk will share some new machine learning methods and systems to profile users, locations, and public transportation services. The reasonably good accuracy of these methods also allow them to be deployed in urban application solutions.

Biography:

Ee-Peng Lim is a professor at the School of Information Systems of Singapore Management University (SMU). He received Ph.D. from the University of Minnesota, Minneapolis. His research interests include social network and web mining, information integration, and digital libraries. He is the co-Director of the Living Analytics Research Center (LARC) jointly established by SMU and Carnegie Mellon University. He is currently an Associate Editor of the ACM Transactions on Information Systems (TOIS),

ACM Transactions on the Web (TWeb), IEEE Transactions on Knowledge and Data Engineering (TKDE), Information Processing and Management (IPM), Social Network Analysis and Mining, Journal of Web Engineering (JWE), IEEE Intelligent Systems, International Journal of Digital Libraries (IJDL) and International Journal of Data Warehousing and Mining (IJDWM). He was a member of the ACM Publications Board until December 2012. He serves on the Steering Committee of the International Conference on Asian Digital Libraries (ICADL), Pacific Asia Conference on Knowledge Discovery and Data Mining (PAKDD), and International Conference on Social Informatics (Socinfo).

2. Prof. T.C. Chang

Title: Critical Tourism Studies in Asia: Issues and Questions

Abstract:

When King and Porananond (2014: 6) first used the phrase "[to] Asianise the field" in their edited book Rethinking Asian Tourism: Cultures, Encounters and Local Response, they were encouraging greater indigenous scholarship in Asian tourism studies. It is their belief that Asian scholars working in and on Asia bring useful insights and unique perspectives that might differ from non-Asian research, thereby creating more holistic understandings of the world. Such an Asian-centric scholarly inclination has bloomed particularly in the 2000s following increasing numbers of Asians on tour, greater intra- and domestic-tourism within Asia, and the opening of new travel destinations throughout the continent. Around the same time as this surge in Asian tourists and indigenous tourism scholarship is an increasing attention paid to 'Critical Tourism Studies' (CTS). The basic premise of CTS is the application of social-cultural theory in tourism analyses, the acknowledgement of researcher bias, and a challenge to mainstream knowledge through a focus on the 'Other' (which might include the marginalised, oppressed and the indigenous among other interest groups). In this presentation, I will consider the 'global' scope of CTS and the 'local' field of Asian tourism knowledge in three ways. Firstly, we will examine what is critical about CTS and how it has evolved over time. Secondly the implications of CTS for Asian tourism research will be explored, and finally issues and questions surrounding critical Asian tourism scholarship will be surfaced. It is the aim of this talk to envision the full potential of CTS if the breadth and depth of 'local/Asian' tourism knowledge is brought to bear on its 'global/lofty' research agenda.

Biography:

T.C. Chang is an Associate Professor at the Department of Geography, National University of Singapore (NUS). He was also Assistant and Vice Dean of the Faculty of Arts and Social Sciences (NUS) from 2008 till 2015. He is the co-editor of two books on Asian tourism: Interconnected Worlds: Tourism in Southeast Asia (with Peggy Teo and K.C. Ho; Elsevier, Oxford) and Asia On Tour: Exploring the Rise of Asian Tourism (with Tim Winter and Peggy Teo; Routledge, London and New York). His research interests include Asian tourism knowledge; urban tourism and urban development issues; arts, culture and creativity.

3. Dr Kao Cai

Title: An Innovative Model for Municipal Solid Waste Collection Routing in Singapore

Abstract: To collect and dispose growing amounts of municipal solid waste (MSW) changed to be a hot topic along with the rapid urbanization in past decades. Cities are more and more dependent on the incineration instead of landfilling due to the cost-efficiency and environmental concerns. Considering the limited number of incineration plants and complicated situation of transportation in both spatial and temporal dimensions in different cities, the optimal routing for waste collection turns to be meaningful research topic. In this research, the ant colony optimization (ACO)-based multi-objective routing model coupled with min-max model and Dijkstra's algorithm is proposed to address the question of which route to take from these waste-generating points to the target incineration plant(s)

IOP Conf. Series: Earth and Environmental Science **38** (2016) 011002 doi:10.1088/1755-1315/38/1/011002

considering travel time, accident probability (black spots), and population exposure, so as to support the routing decision-making. The model is successfully implemented in Singapore and the effectiveness of the model has also been justified. Besides, few limitations of this research have also been discussed, some of which would also be the future directions of our research, especially the design and integration of a web-based routing decision-making support system.

Short Bio:
Dr. Kai Cao holds a Ph.D. degree in Geography from the Chinese University of Hong Kong in Hong Kong and obtained his B.S. and M.Phil. degrees in Geography from Nanjing University in China. Prior to joining Department of Geography at NUS, he had also worked in Center for Geographic Analysis at Harvard University, Department of Geography at University of Illinois at Urbana–Champaign, and World History Center at University of Pittsburgh respectively. His current research interests center on GIScience and its applications, especially on the topics of spatial simulation and optimization, cyberinfrastructure and geocomputation, land use planning support and urban mobility.

4. Prof. Ida Ayu Dwi Giriantari

Title: Renewable Energy Deployment in Indonesia; opportunity and challenge

Short Bio:
Ida Ayu Dwi Giriantari is Professor in Electrical Engineering Udayana University. She hold PhD from The University of New South Wales in 2003. Her research areas are in Power Plant, Renewable Energy and Power Transformers. She is Head of Magister Program of Electrical Engineering, Udayana University.

IOP Conf. Series: Earth and Environmental Science **38** (2016) 011003 doi:10.1088/1755-1315/38/1/011003

Peer review statement

All papers published in this volume of *IOP Conference Series: Earth and Environmental Science* have been peer reviewed through processes administered by the proceedings Editors. Reviews were conducted by expert referees to the professional and scientific standards expected of a proceedings journal published by IOP Publishing.

Seismic Data Analysis to the Converted Wave Acquisition: A Case Study in Offshore Malaysia

A.H. Abdul Latiff[1], S.A.A. Osman[2], S.N.F. Jamaludin[3]

[1, 2, 3] Centre for Seismic Imaging, Geosciences Department, Universiti Teknologi PETRONAS, Malaysia

Email: abdulhalim.alatiff@petronas.com.my

Abstract: Many fields in offshore Malaysia suffer from the presence of shallow gas cloud which is one of the major issues in the basin. Seismic images underneath the gas cloud often show poor resolution which makes the geophysical and geological interpretation difficult. This effect can be noticed from the amplitude dimming, loss of high-frequency energy, and phase distortion. In this work, the subsurface will be analyzed through the geophysical interpretation of the converted P-S data. This P-S converted dataset was obtained through ocean bottom cable (OBC) procedure which was conducted at a shallow gas affected field located in Malaysian Basin. The geophysical interpretation process begin by picking the clear faults system and horizons, followed by thorough post-stack seismic data processing procedure. Finally, the attributes analyses were implemented to the seismic section in order to image the unseen faults system. The interpreted seismic sections show significant improvement in the seismic images, particularly through median filter process. Moreover, the combination of structural smoothing and variance procedure had contributed to the correct faults location interpretation.

Keywords: converted wave, attribute analysis, Malaysia basin

1. Introduction

The potential hydrocarbon reservoir in Malaysia basin is always associated with the presence of shallow subsurface anomaly such as carbonate platform and shallow gas accumulation. While imaging techniques beneath carbonate body are well studies, a good quality seismic image underneath the shallow gas cloud remained a challenge in geophysics community. By definition, shallow gas cloud is an overburden region of low-concentration gas, escaping and migrating upward from a gas accumulation [1]. Generally it shows as a region of deteriorated seismic data quality associated with slow wave propagation velocity and reflection time sag beneath the gas zone. The P-wave reflected events in this region appear with lower amplitude and frequency content, which is due to wave energy's anelastic and intrinsic loss. Such undesired effects lead to significant uncertainties in structural and stratigraphic mapping of reservoirs below the shallow gas area, and consequently the hydrocarbon resource assessment of the affected fields. This will affect the geologist and geophysicist's ability in predicting and interpreting the structures and reservoir properties beneath shallow gas cloud (Figure 1).

[1] Corresponding author

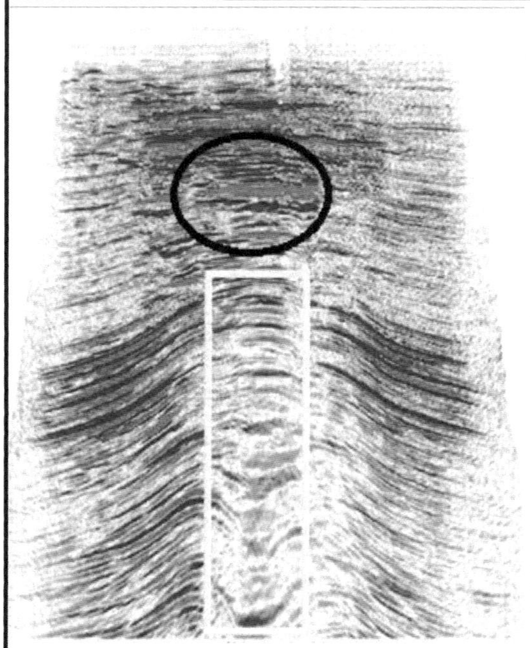

Figure 1: The stack section from Malaysia basin shows a shallow gas accumulation (black circle) which affects the reflectors data quality underneath the gas zone (yellow box). [2]

To overcome the poor data quality produce by P-wave seismic acquisition in shallow gas affected area, current industry practice use the ocean bottom cable (OBC) technology that recorded reflected shear waves signal, which include larger signal frequency bandwidth that has no offset limitations. In addition, the converted shear waves generated from sea floor are unaffected by the presence of gas accumulation, which can give good reflections images that not visible on compressional wave data. Therefore, by using converted shear waves obtained from (OBC) technology, the shallow gas cloud effects can be minimized. On the other hand, the downside of this technology is that it only applicable to water-depth below 200m as well as the high cost of OBC vessel mobilization and demobilization. In this study, we evaluated the P-S converted waves obtained from Malaysia offshore in term of geophysical approach such as faults and horizons picking, implementation of post-stack processes and utilization of seismic attributes. At the end of this work, the outcomes generated from this work will provide some guidelines on optimum imaging of the structures at and beneath the shallow gas clouds.

2. Imaging Issues

The shallow gas cloud was formed due to the gas leakage through sediments. It is believed that faults or fractures are the main pathway for the gas migration towards shallower unconsolidated sediments [3]. There are three main indicators of shallow gas cloud; seismic acoustic turbidity, high amplitude reflection at the top of the gas zone and acoustic blanking at the bottom of the gas zone [4]. The strong reflected energy was recorded at the top of gas zone as the large acoustic impedance contrast was formed between the gas-filled with finely-layered porous silt-sand-rich, and the clay-sand-rich clastic sediments. Typically, a high portion of the reflected acoustic energy is reflected to the surface, leaving a smaller amount of transmitted energy through the gas zone. Therefore, this reflection response contains information on the properties of the shallow gas body. Due to heterogeneous nature of shallow gas-filled zone, it resulted in scattering, dispersion, internal multiples, wave mode conversion as well as anelastic losses. Due to strong acoustic impedance contrasts and the low velocity of the gas layers, the field shot gathers will show strong internal short-period multiples and non-hyperbolic moveout.

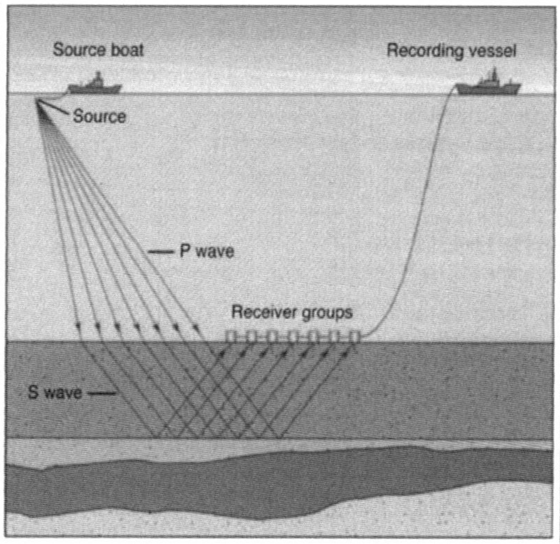

Figure 2: The figure shows ocean bottom cable arrangement in offshore acquisition. [5]

During the last few years, a number of 3D OBC surveys have been acquired all over the world as a countermeasure to conventional towed streamer technique. OBC technology is a hybrid marine and land seismic data acquisition operation where the system deploys a stationary array of multi-components receiver devices on the ocean floor while the marine vessel produce the seismic energy source (Figure 2). In conventional towed streamer, seismic data were processed and imaged only for the compressional wave (P-wave) reflections. This indicate that the P-wave propagates downward from the seismic source before being reflected and propagate upward to the receivers, also as P-wave. In the case of OBC data, converted shear wave data were recorded on the horizontal receiver components available from the multi-components receivers located on the sea floor. This converted shear wave was produced once the down-going P-wave partitioned its energy into reflected and refracted P and S-waves.

Due to poor seismic imaging and the fact that there are only a few available wells that had been penetrated beyond the gas accumulation zone, there is no in-depth knowledge as to the actual geology within and beyond the shallow gas cloud. However, the presence of shallow gas cloud leads to poor interpretation of the structures beneath this particular region. The shear wave imaging has several notable advantage compare to compressional wave imaging. As the wave propagate through the rock-gas matrix, shear wave is unaffected by the compressibility of the gas as compare to other wave types. In addition, the usage of S-wave imaging has contribute to:

a. Higher signal-to-noise ratio.

b. Broader signal bandwidth, which consequently improves data resolution.

c. Seismic data multiple elimination. This can be achieved by measure both pressure and particle motion, thus separate the up going wave from the down going wave.

d. Applicable for wide azimuth acquisition.

e. Quieter recording environment.

3. Results and Analysis

The converted shear wave, P-S data was analysed and compared to the conventional P-P seismic data in term of the seismic data quality. As expected, the P-S data shows a significant enhancement to the seismic images, as important reflectors clearly visible, while less acoustic turbidity has been recorded (Figure 3). The gas cloud effects which appear in the conventional data is no longer visible in the P-S data. However, the shallow gas zone could still be recognised in the P-S data.

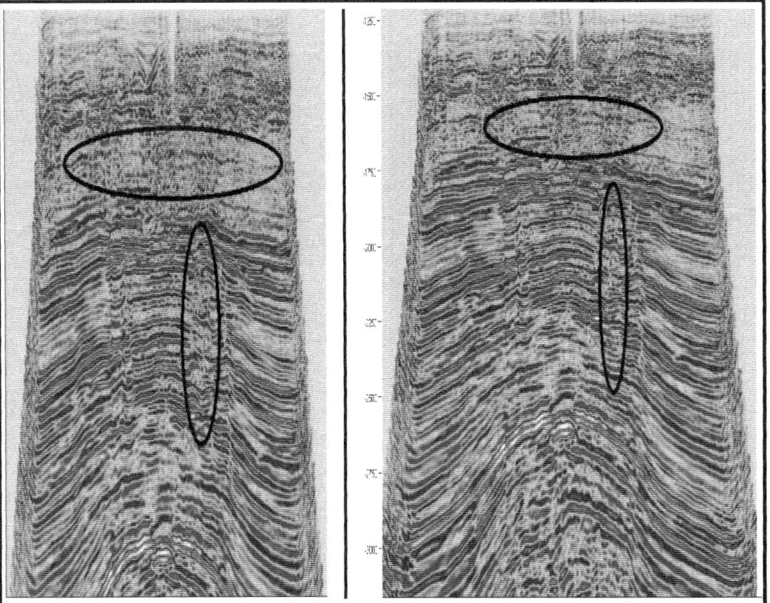

Figure 3: The figures show the comparison between P-P wave imaging (a) and P-S wave imaging (b). Both imaging techniques still suffer from shallow gas cloud issue, as being highlighted by black ellipsoid.

3.1 Faults and Horizons Interpretation

The structural framework was implemented by picking assigned fault segments on inline sections of seismic with the trace appearing on the corresponding cross lines. These faults are represented on the seismic sections as a discontinuous reflection along a preferred orientation of reflectors or as distortion of amplitude around the fault zones. A total of two faults named as F1, F2, were identified (Figure 4). However, many difficulties have been encountered during picking the faults. Uncertainties during picking the correct position and extension of the faults were the main issues in this part. The shallow section of the seismic image still suffers from the acoustic turbidity, thus the extension of the faults cannot be distinguished.

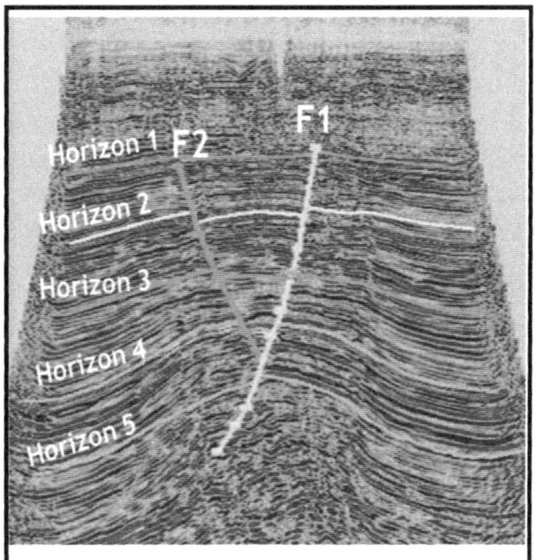

Figure 4: The inline section shows five seismic horizons and two fault systems picked in this area.

Horizons picking are significant in order to get a clear picture on what have been observed from the shallow gas cloud effects. A total number of five horizons were interpreted from the data with time maps generated for each of the horizon. Some of the horizons were partially affected by the shallow gas cloud effect, while others were significantly affected specifically by the amplitude dimming. This in turn has led to some ambiguity while picking the horizons. The gas trapped in the shallow section attenuated the seismic signal in the area beneath it. Consequently, the image quality is deteriorated with lower resolution, amplitude and frequency, making the interpretations and horizons picking difficult.

3.2 Post Stack Processing

The primary objective of implementing the post-stack processing is to enhance the data quality through the gas cloud zone. Several processes were applied such as first derivative, frequency filter, median filter, original amplitude, time gain, and trace gradient. However, only two processes have provided a significant enhancement of the images; time gain and median filter, which able to reduce the effect of acoustic turbidity. The median filter is a non-linear digital filtering technique, often used to remove noise. The main idea of the median filter is to run through the signal entry by entry, replacing each entry with the median of neighboring entries. The pattern of neighbors is called the "window", which slides, entry by entry, over the entire signal. As a result, the seismic images showed a significant reduction in the acoustic turbidity (Figure 5).

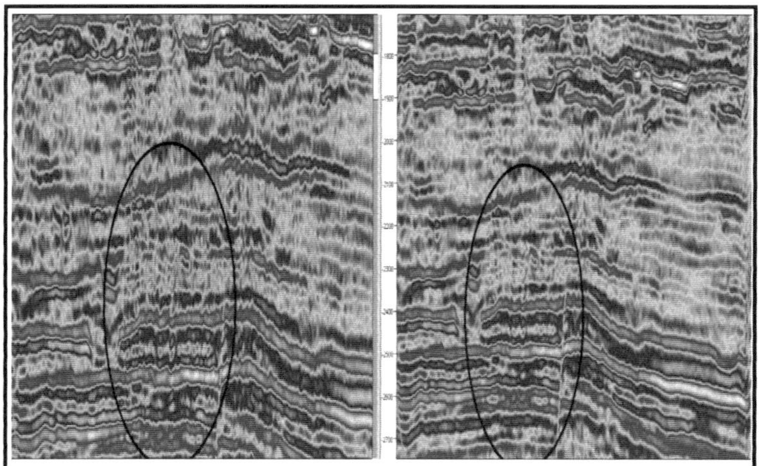
Figure 5: The inline sections show the outcomes from median filter (b), compare to original seismic data (a). The black ellipsoids show the region of interest.

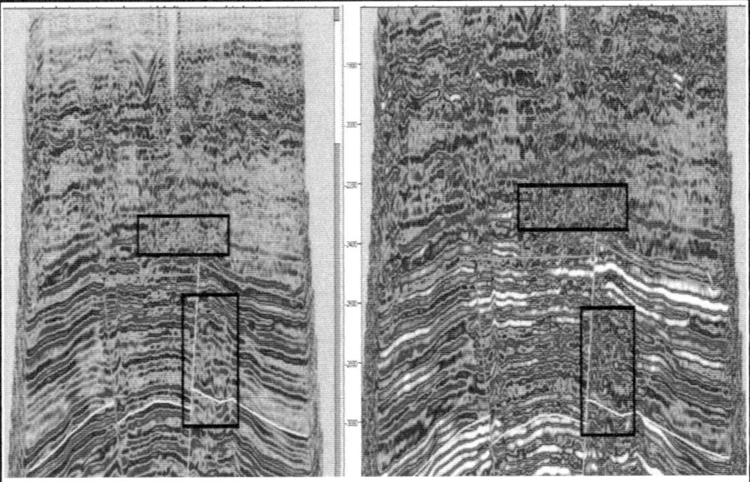

Figure 6: The inline sections show the outcomes from the time gain (b) compare to the data after median filter (a). The black boxes show the region of interest.

The median filtering technique is also one kind of smoothing technique, which is effective at removing noise in smooth patches or smooth regions of a signal. However, at the same time, the process adversely affects edges. The correction for the loss in amplitude of the signals is required in order to enhance the overall seismic images. In this case, the time gain procedure was applied, which involves multiplication of the signal by a number that increases with time. The outcome reveals that the amplitude and frequency losses are partially recovered in the areas where gas clouds present. The resulting image has improved resolution, horizons continuity, thus have improved the interpretations. The last step in post-stack processes was to apply time gain and median filter to the same seismic section in order to optimize the overall image (Figure 6). It is noticed that the true position of the fault can be clearly picked as pointed by the black lines. As a result, surface horizon maps were generated to correct our horizons interpretation.

3.3 Attributes Analysis

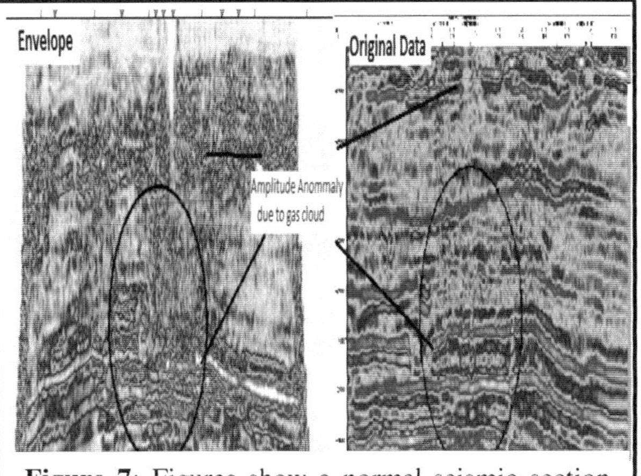

Figure 7: Figures show a normal seismic section that contain a high amplitude anomaly at shallow depth due to the presence of shallow gas cloud (b), thus caused high amplitude anomaly in envelope.

Seismic attributes form an integral part of qualitative interpretative tool that facilitates structural and stratigraphic features (e.g. channels, pinch out and meanders) as well as offer clues to lithology type and fluid content estimation with a potential benefit of detailed reservoir characterization. Seismic attributes such as instantaneous phase, envelope, structural smoothing, local structural azimuth, local structural dip, and variance were applied based on volume interpretation. The objective of implementing these attribute methods is to improve the faults' interpretation. The first implemented attribute was the envelope which represents the instantaneous energy of the signal and is proportional in its magnitude to the reflection coefficient. It represents the acoustic impedance contrast, which is useful in highlighting gas accumulation (Figure 7), discontinuities, faults, tuning effect, and sequence boundaries. It is proportional to reflectivity and therefore useful for analyzing AVO anomalies. If there are two volumes that differ by constant phase shift only, their envelopes will be the same.

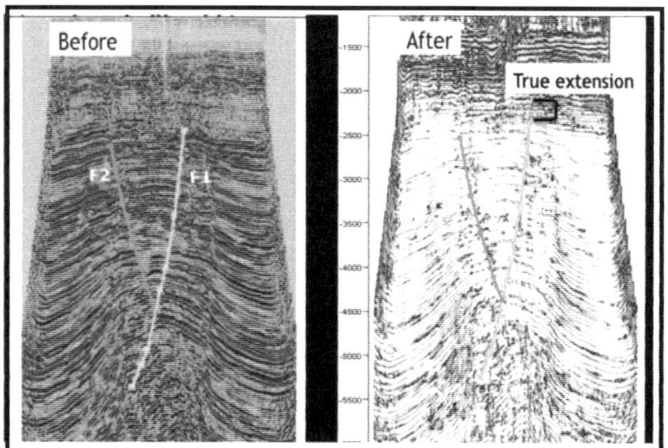

Figure 8: Figures show the application of variance attribute: (a) after post-stack analysis, (b) after application on variance and structural smoothing attributes.

The next step in implementing seismic attributes is to enhance our faults interpretation. The extracted attributes were analyzed and it was observed that only variance, combined with structural smoothing, showed better response to faults than any other attribute in the gas cloud area. For this study, the variance attribute was calculated by allowing the software to perform structural smoothing attribute (Figure 8). Subsequently, the results from variance and structural smoothing attributes show that, F1 fault system did not terminate as previously interpreted. Therefore, the interpretation of faults has been improved, in which the true extension of the fault was observed as shown by the black brackets.

4. Conclusion

To sum up, the primary objective of this work is to analyze and enhance the seismic images beneath the shallow gas cloud areas for a field located in the Malay Basin. From the results, it show that post-stack processing methods able to improve the seismic images. This was clearly seen by the improvement in horizons visibility, as well as acoustic turbidity reduction. Meanwhile, seismic attributes also had given a better way to detect the true position of faults, by combining structural smoothing and variance. Lastly but not least, the results of geophysical analysis are very encouraging, however, further analysis is necessary to fully optimize the seismic images of this dataset.

Acknowledgement

We would like to thank PETRONAS's Malaysian Petroleum Management (MPM) for providing the seismic data and UTP for funding this research project (0153AA-D70).

References

[1] R. E. Sheriff, "Encyclopedic dictionary of exploration geophysics," *Society of Exploration Geophysicists (SEG)*, 2002.
[2] M. Madon, "The petroleum geology and resources of Malaysia," *PETRONAS*, 1999.
[3] A. R. Ghazali, "True-amplitude seismic imaging beneath gas clouds," *PhD, Applied Physics, Delft University of Technology*, 2011.
[4] A. H. Abdul Latiff, D. P. Ghosh, and S. N. F. Jamaludin, "Structural Enhancement in Shallow Gas Cloud Region," *IOP Conference Series: Earth and Environmental Science*, vol. 23, p. 012005, 2015.
[5] http://www.peakseismic.com/content/ocean-bottom-seismic.asp

Enhancement in Seismic Imaging using Diffraction Studies and Hybrid Traveltime Technique for PSDM

Y. Bashir[1, 2], D. P. Ghosh[1], S. Y. Moussavi Alashloo[1], and C. W. Sum[1]
[1] Centre for Seismic Imaging (CSI), Department of Petroleum Geoscience, Universiti Teknologi PETRONAS, Tronoh, Perak, Malaysia

E-mail: Yasir.bashir@petronas.com.my

Abstract. The accurate migration of seismic data is conditional on the parameters which are nominated. The effective velocity used in residual processing for migration is small compared to the original migration velocity. Considering traveltime computation is a significant part of seismic imaging algorithms. Conventional implementation of Kirchhoff migration is essential for precomputing a traveltime table from the categories involving traditional ray-tracing methods and finite difference eikonal solvers. In this paper, we examine the accuracy using, the eikonal solver and paraxial ray tracing traveltime computation in pre-stack Kirchhoff depth migration. This hybrid traveltime technique can be applied to a variety of problems related to faults, fractures, and complex region. To evaluate the relevance of this identical traveltime technique, we applied on a Marmousi data set.

1. Introduction

The expectation of a geophysicist about the subsurface is as complex as he/she can deliberate; this is, because real geology is undergone with dissimilar procedures and a long tenor to survive. So, information approached through seismic imaging will be more complex as the velocity is a multifarious structure. Marmousi is one of the complex and standard model which have been used to prove seismic imaging algorithms. The downward continuation of source and receiver by finite difference is relatively slow and dip-limited compared to the Kirchhoff integral implements migration, which has no dip inadequacy[1].

Migration is the one of primary seismic imaging tools; but, the earliest analogue seismic record was an illustration of a simple single-fold. These records were comprised of diffracted energy and random noise; but, those records still gave an interpretation of the earth's subsurface. Mechanical migration removed the structural misrepresentation on early seismic data, and the CMP stack condensed the amount of random noise when diffracted energy was preserved. The subsequent development of DMO [2] was used a method to expand the dip bandwidth of CMP stacking. Later on, the discovery of DMO (that is a procedure of pre-stack migration), led to a demand for migration methods with greater steep-dip imaging together with depth migration, leading to the wide range of migration techniques currently available [3].

2 Yasir Bashir (Yasir.bashir@live.com)

Traveltime computation plays an important role in seismic imaging. Several migration algorithms have been established on ray theory (high-frequency assumption). The lesser the curvature or the greater the radius of the curvature will give an improved quality of the ray-path. On the other hand, if the curvature of the ray is too big, the projecting point is not single, and also the Taylor series expansion is inaccurate; so, a poor migration result may suffer from the inaccurate calculation of traveltime [4].

In this paper, we have used a processing workflow including sorting into CMP order, NMO correction and stacking for diffraction studies. Using slope frequency filtering in f-k domain, separation of diffracted event and dipping reflector is done for diffraction studies. Kirchhoff PSDM is applied to the pre-stack data for imaging and traveltime is computed by paraxial ray-tracing and Eikonal solver which is a hybrid traveltime method for enhancement of imaging results.

2. Methodology

2.1. Theory and Method

Finite Difference Method

Here, for the modelling process, we took the velocity and density as the input, and the output was the seismic data. For the seismic inversion case, the input is traces and the out is the structural image. Both of these (Modelling and Inversion), common quantities are required to be calculated. These quantities consist of [4]:

$\emptyset$ Propagation Angle
τ Travel Time
β Incident Angle from Source or Receiver
σ Running Ray Parameter
$\partial\beta/\partial x$ Geometrical Spreading Parameter

For each source or receiver, the above quantities satisfied the following equation[4]:

$$\left(\frac{\partial\tau}{\partial x}\right)^2 + \left(\frac{d\tau}{dx}\right)^2 = \frac{1}{v^2(x,z)} \qquad (1)$$

$$\sin\emptyset = v\frac{d\tau}{dx} \qquad (2)$$

$$\frac{\partial\sigma}{\partial x}\frac{\partial\tau}{\partial x} + \frac{\partial\sigma}{\partial z}\frac{\partial\tau}{\partial z} = 1 \qquad (3)$$

$$\frac{\partial\beta}{\partial x}\frac{\partial\tau}{\partial x} + \frac{\partial\beta}{\partial z}\frac{\partial\tau}{\partial z} = 1 \qquad (4)$$

$$\frac{\partial}{\partial z}\left(\frac{\partial\beta}{\partial x}\right) + \frac{\partial}{\partial x}\left[\mu(x,z)\frac{\partial\beta}{\partial x}\right] = 0 \qquad (5)$$

where v(x, z) is the velocity and

$$\mu(x,z) = \frac{\partial\tau}{\partial x}\left[\frac{\partial\tau}{\partial z}\right]^{-1}$$

*E*quation (1) eikonal equation, Equations (3) and (4) are derived by Pusey and Vidale (1991). Equation (5) follows from equation (3).

2.2. Travel time computation

Figure 1 describes the ray-tracing phenomena in a complex and heterogeneous medium with 100 rays. Paraxial optics can be defined as a ray-tracing accomplished in the limits of very small ray angles and heights. They allow us to make a number of simplified assumptions that make the arithmetic of ray-tracing considerably easier.

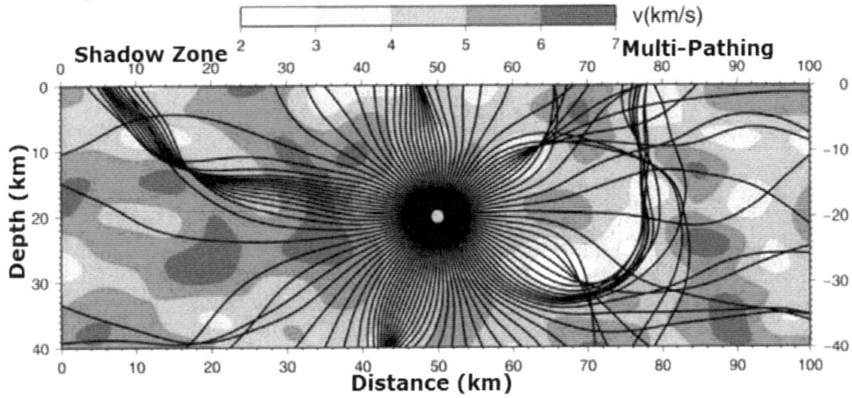

Figure 1: Trajectories of 100 rays emitted by a source point in a heterogeneous medium. Shadow zone and multi-pathing effect can be seen [5].

The eikonal differential equation is used for the elementary mathematical models which describe the traveltime (eikonal) propagation in a velocity model. The benefits of this method in comparison with ray tracing techniques comprise the capability to work on consistent model grids, a wide-ranging coverage of the receiver space, and a reasonable numerical robustness. A common implementation of the finite-difference eikonal equation is calculating the first-arrival travel times, though frequency-dependent enrichments.

The eikonal equation, describing the traveltime propagation in an isotropic medium, has the form:

$$(\nabla\tau)^2 = n^2(x, y, z)$$

where $\tau(x, y, z)$ is the traveltime from the source to the receiver point with the coordinates (x,y,z), and n is the slowness at that point (the velocity v equal to 1/n) [6].

This method calculates each traveltime table by paraxial ray-tracing, then in the shadow zone (Figure 1) the traveltimes are computed by solving the eikonal equation.

3. Results and Discussion

We have tested the above travel time calculation method to validate the results in a 2D Marmousi model. The Marmousi model has been considered as identical with the phrase complex media (Figure 2). The huge numbers of faults and folds induced in the model has created an interesting distribution of velocity anomalies and discontinuities. Thus, the Marmousi model has serve as a calibration tool used to test the various traveltime and migration algorithms. The data was produced using a 2-D acoustic finite-difference modelling programme.

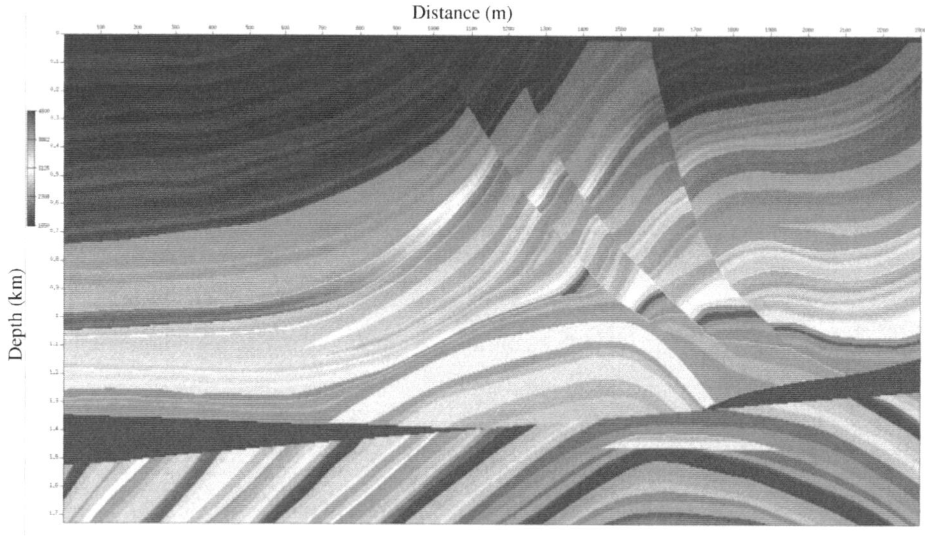

Figure 2: Marmousi model, velocity range from 1650 m/sec to 4600 m/sec. 3 major faults, unconformity, anticline structure and overburden in the deeper section).

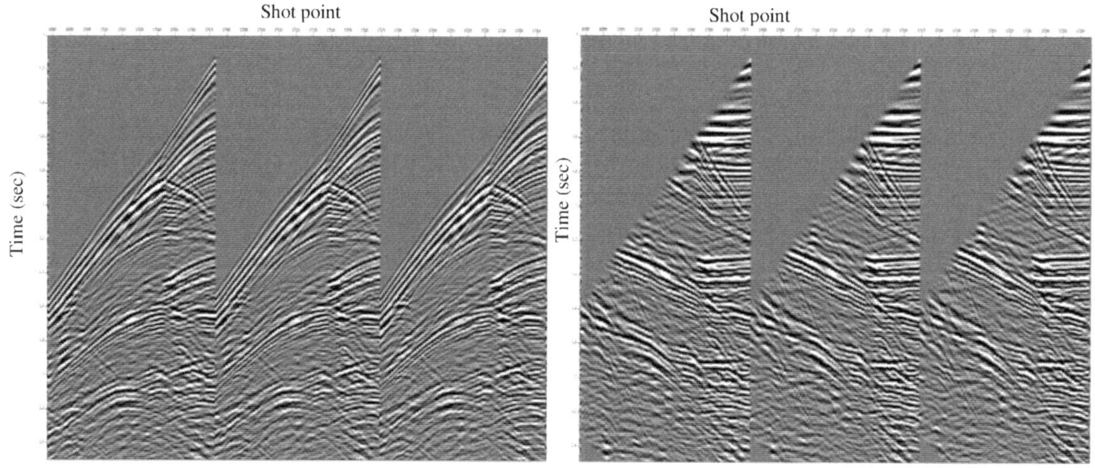

Figure 3: Seismic gathers before the NMO (Normal move out) correction (left). Seismic gather after the NMO correction. Events are slightly strait at the true reflection points (right).

Velocity analysis is a crucial job in processing; before the NMO correction we picked the correct velocity at different points. Figure 3 (left), we can see the move-out decreased when an increasing time and velocity; this is cause of poor resolution in deeper area. Figure 3 (right) shows selected shot points (17070, 17170, & 17270) from the section to illustrate the results after the NMO correction.

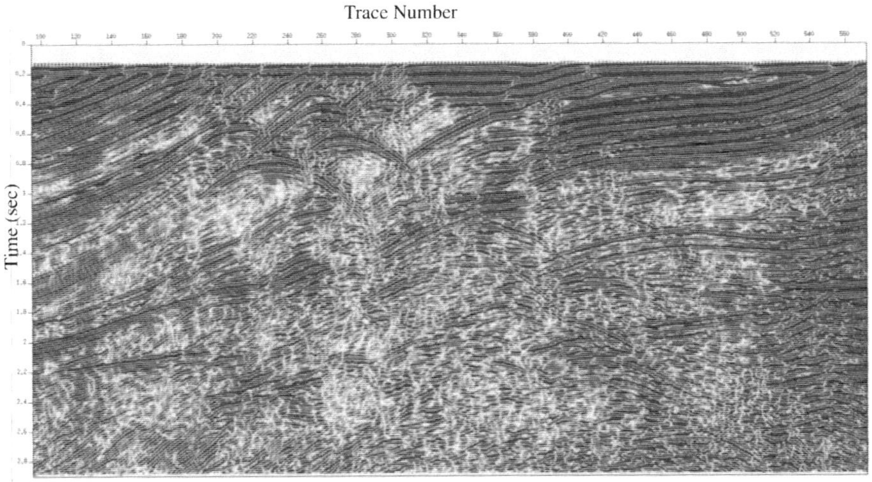

Figure 4: Stacked seismic sections after sorting the gathers to the CMP. Diffraction hyperbola can be seen at the faults, pinchouts and discontinuity.

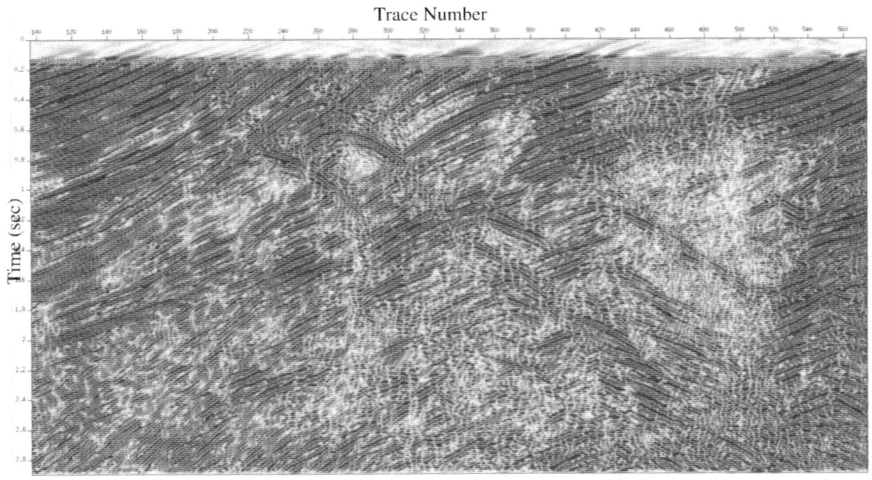

Figure 5: Seismic Diffraction section by using the linear interpolation technique to determine the slope and amplitude on the dipping reflectors.

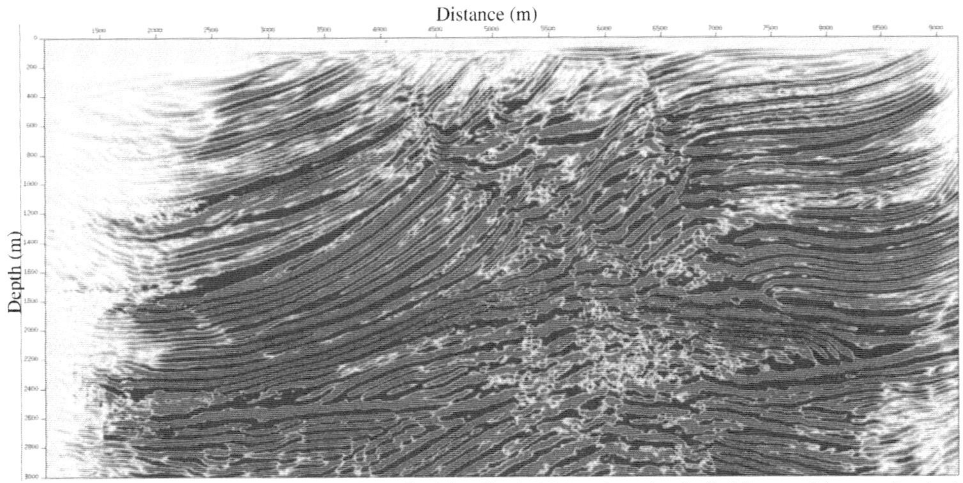

Figure 6: Seismic sections after the Kirchhoff PSDM using the hybrid traveltime calculation technique.

IOP Conf. Series: Earth and Environmental Science **38** (2016) 012002 doi:10.1088/1755-1315/38/1/012002

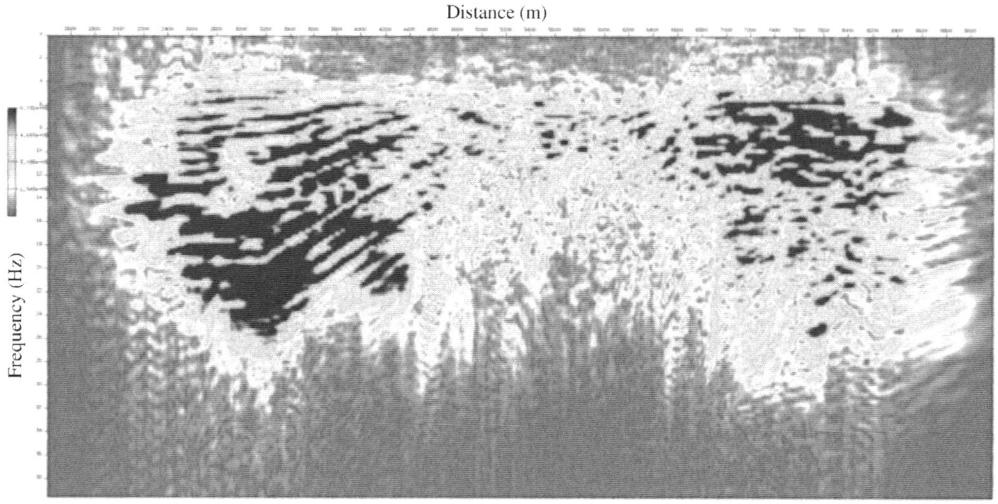

Figure 7: Amplitude spectrum of the migrated data

Figure 4 shows the stacked section before migration, in which we can recognise the diffraction hyperbolas where the faults and pinchouts are in the model (*Figure 2*). Migration aperture is one of parameters to stack the diffraction hyperbola. In this algorithm, we have used the 40 degree migration aperture to stack the hyperbola energy at the apex. Migration aperture smaller than the optimal size, will cause the destruction of the faults, fractures and pinchouts. More importantly, large apertures will degrade the migration quality in a poor signal-to-noise ratio and increase the cost of the computational power without any quality improvement [7]. After converting the shot gather to the CDP gather we applied the frequency dip filtering in the F-k domain to separate the diffraction and faults from the reflection (*Figure 5*). *Figure 6* shows the final results of the Kirchhoff migration using a hybrid traveltime. Faults and pinchouts that make a complex velocity structure in deeper sections have been illuminated and can be interpreted properly. Amplitude spectrum is achieved by applying Fourier transform to the data in time domain (*Figure 7*), which is explaining the amplitude distribution with frequency and trace number. We can see the amplitude in the deeper section is affected by the faults and cause of poor amplitude recovery in the deeper section, both side on the corners amplitude is recovered.

4. Conclusions

We integrated two approaches to illuminate the subsurface structure using initial processing and migration. A hybrid traveltime computation method for PSDM has been studied in the imaging context. The propose study has validated that, the hybrid traveltime computation method for Kirchhoff migration is a tool for enhancement of imaging results in complex geological settings. In conventional data processing, either the eikonal solver or paraxial ray-tracing are used. The results demonstrate the significance of using hybrid traveltime computation, and provide more conceding results. We have used a Marmousi data set to prove this application algorithm, which is a complex geological model. On the processing side, sorting, CMP gathering, NMO correction, and stacking have been performed for the diffraction studies. Diffractions have been separated using dip frequency filtering in the f-k domain. The existence of a hyperbolic pattern is an indication of faults and/or fracture because of abrupt changes of velocity and acoustic impedance contrast. After application of the imaging algorithm with the correct aperture selection, each hyperbola was stacked at their apex. Overall, the hybrid traveltime imaging algorithm enables one to improve the imaging result for complex structures such as faults, fractures and structures below the complex velocity.

Acknowledgments

We are thankful to Universiti Teknologi PETRONAS (UTP) and the Centre for Seismic Imaging (CSI) to provide the facility for this research. Authors are thankful to UTP & PETRONAS staff for their contribution. In this work, we have used Seismic Unix from CSM.

Reference

[1] Z. Liu, "A Kirchhoff approach to seismic modeling and prestack depth migration," in *SEG Technical Program Expanded Abstracts 1993*, pp. 888–891.
[2] Ö. g ˇ an Yilmaz and J. F. Claerbout, "Prestack partial migration," *Geophysics*, vol. 45, no. 12, pp. 1753–1779, 1980.
[3] S. H. Gray, Etgen, John;, Dellinger, Joe;, and D. Whitmore, "Seismic migration problems and solutions," *Geophysic*, vol. 66, no. 5, p. 1622.
[4] N. Bleistein, "TWO-AND-ONE-HALF DIMENSIONAL IN-PLANE WAVE PROPAGATION*," *Geophys. Prospect.*, vol. 34, no. 5, pp. 686–703, 1986.
[5] N. Rawlinson, J. Hauser, and M. Sambridge, "Seismic ray tracing and wavefront tracking in laterally heterogeneous media," *Adv. Geophys.*, vol. 49, p. 203, 2008.
[6] S. Fomel, "Traveltime computation with the linearized eikonal equation," *SEP*, vol. 94, pp. 123–131, 1997.
[7] Y. Bashir, D. Ghosh, and C. W. Sum, "Detection of Fault and Fracture using Seismic Diffraction and Behavior of Diffraction Hyperbola with Velocity and Time," in *Seismic Driven Reservoir Characterization and Production Management*, 2015.

Comparison of two typhoon-induced storm surges at the Zhanjiang Coast

Xue Zhao[1], Qiang Xie[2], Bo Hong[3*], Hongzhou Xu[2], Lingfang Chen[2], Jinlei Hu[1], Min Zhang[1], Baoxia Huang[1], Zhiwei Che[4]**

[1]South China Sea Forecast Center, State Oceanic Administration, Guangzhou, Guangdong 510310, China

[2]Institute of Deep-sea Science and Engineering, Chinese Academy of Sciences, Sanya, Hainan 572000, China

[3]School of Civil and Transportation Engineering, South China University of Technology, Guangzhou, Guangdong 510641, China

[4]Haikou Marine Environment Monitoring Center Station, State Oceanic Administration, Haikou, Hainan570311, China

*Equal to the first author
**Corresponding author: hzxu@sidsse.ac.cn

Abstract. Typhoon-induced storm surge is the largest threat to the Zhanjiang Coast of the northern South China Sea. In this study, we report on two processes of storm surges generated by Typhoons Rammasun (TR) and Kalmaegi (TK) with similar tracks at this region. Results show that the peak of the storm surge during TK (with a maximum value of 4.50 m) was significantly larger than that during TR (with the maximum value of 2.60 m), although the intensity of TK (Category 4 with 40 m s^{-1} maximum wind and 960 hPa central air pressure) was much weaker than that of TR (Category 6 with 60 m s^{-1} maximum wind and 910 hPa central air pressure). Comparisons of the typhoon properties and astronomic tides reveal that the peaks of storm surge were closely related to typhoon maximum wind radius, typhoon moving speed, and tide-surge interaction rather than typhoon intensity during the two cases, indicating complex dynamics of storm surge at the Zhanjiang Coast.
Keywords: Storm surge, Typhoon, Wind radius, Moving speed, Zhanjiang Coast

1. Introduction

The storm surge hazard induced by landing typhoons in Asia and hurricanes in North America (both are named typhoon in this study) has become a serious threat to the coastal ecological environment and its dynamics have been studied for several decades. Weisberg and Zheng (2006) pointed out that the intensity of a typhoon is a key factor for determining the strength of the storm surge in the Tampa Bay off the Florida coast based on numerical simulations. Their result indicated the height of the storm surge was proportional to the square of the wind speed in the Tampa Bay. Powell and Reihold (2007) also suggested storm surge can be estimated by kinetic energy of a typhoon based on the analyses of the relation between storm surge and typhoon kinetic energy during several typhoons in North America.

Proudman (1953) concluded that the moving speed of a typhoon has an important impact on the peak of the storm surge, in which storm surge reached maximum value when the moving speed was close to the propagation speed of its long wave. Rego and Li (2009) studied the responses of storm surge to hurricanes with different moving speeds at the coasts of Louisiana and pointed out that typhoons with faster moving speeds can indeed cause higher surges, but with faster dissipation rates. A similar phenomenon has also been found in the Florida Bay (Zhang et al., 2012). However, the peaks of storm surges are reduced by typhoons with faster moving speeds in some areas with complex topography, such as the Tampa Bay in Florida (Weisberg and Zheng, 2006), North Carolina coasts (Peng et al., 2004), and Hangzhou Bay (Han and Pan, 2013), where the propagation speed of long waves is smaller than the moving speed of the typhoons. Xu et al. (2010) found the peaks of storm surges were determined by typhoon tracks around the Florida coasts and found that topography with half-closed bay and estuary more easily allowed for accumulation of the water column. Han and Pan (2013) found different typhoon tracks caused largely different storm surges at the Hangzhou Bay. In addition, the shallow continental shelf is favorable for formation of a storm surge, as the surge wave is upraised suddenly when it runs up against the shallow topography (Rego and Li, 2010). The above studies indicate that the peak of a typhoon-induced storm surge is dependent on a number of factors and the dynamics are very complex.

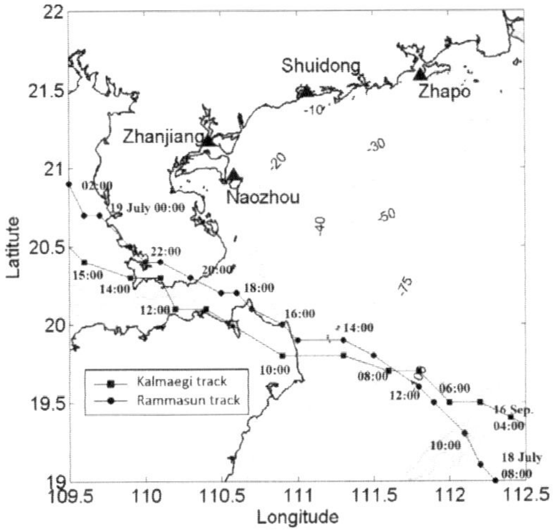

Figure 1. Tracks of TK (squared line) and TR (dotted line). Triangles represent tide-gauge stations at Naozhou, Zhanjiang, Shuidong and Zhapo. Black and gray lines represent coastline and isobaths at the Zhanjiang Coast, respectively.

The Zhanjiang Coast is located on the west coast of Guangdong Province (Figure 1). It connects with the northern South China Sea through a wide and shallow shelf which extends about 230 km from the coastline to 100 m water depth. Its shallow and half-closed shelf makes the Zhanjiang Coast one of the most storm-surge prone areas along China's coastline (Elliott et al., 2015). Dong et al. (2014) studied the temporal and spatial distributions of storm surges along the Guangdong Coasts using a 50-year dataset at ten tide gauge stations and concluded that the Zhanjiang Coast was the most severely affected area with the highest frequency of storm surges. There is no doubt that different typhoons can cause different levels of storm surge along the Zhanjiang Coast. For example, Typhoon 8007, which was a Category 4 storm, caused a 5.90 m storm surge at Nandu tide gauge station along the Zhanjiang Coast, while Typhoon Krovanh, another Category 4 storm, and Typhoon 6508, rated a Category 6, only caused about 3.80 m and 2.40 m storm surges at the same place, respectively (Chen et al., 2002; Shi et al., 2004; Cao et al., 2006). All of the above studies suggest that the height of storm surge was closely related to typhoon intensity, track, and landing location along the Zhanjiang Coast. In this study, we report storm surge processes during two different typhoons with similar tracks and landing locations, but with different intensity, and discuss the potential dynamic mechanisms for the processes at this region. The flow chart of research methodology is showed in Figure 2.

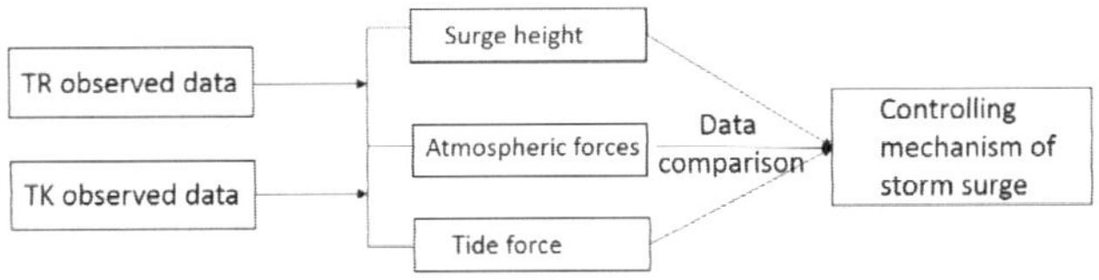

Figure 2. Flow chart of research methodology.

2. Typhoons Rammasun and Kalmaegi

Typhoon Rammasun (TR) initially formed as a tropical depression at about 210 km west of Guam in the Northwestern Pacific at noon of 12 July, 2014. It moved westwards and landed in the central coast of the Philippines after intensifying as a severe typhoon on 15 July. TR entered the South China Sea on 16 July and developed as a super typhoon with maximum 55 m s^{-1} sustained winds and 930 hPa air pressure on 18 July. It moved northwestward and made landfall near Wenchang over the northern Hainan Island at 15:00 on 18 July, becoming the strongest landing typhoon (with 60 m s^{-1} maximum wind and 910 hPa central air pressure) in China since 1949. Seven hours later, it crossed over the Leizhou Peninsula and entered into the Beibu Gulf. It continued to move northward and crossed over Guangxi on 19 July, after weakening to a Category 2 typhoon.

Figure 3. Damages and huge storm surge caused by TR (a) and TK (b), respectively.

Typhoon Kalmaegi (TK) initially generated as a tropical depression about 1100 km southeast of Manila in the Philippines at 14:00 on 12 September, 2014. It rapidly developed into a tropic storm, moving northwest, and then strengthened into a typhoon with 30 m s^{-1} peak winds and 980 hPa air pressure at about 7:00 on 13 September. TK crossed over the Luzon Island and entered into the South China Sea at midnight on 15 September. The intensity of TK remained at almost the same level, with 40 m s^{-1} maximal winds and 960 hPa air pressure, as it drifted northwestward toward the Hainan Island. Like TR, TK also made landfall near Wenchang at 10:00 on 16 September and then moved northwestward across the Leizhou Peninsula and Beibu Gulf. It finally landed at Hatrang in Vietnam at 23:00 on 16 September.

It can be seen that TR and TK had similar tracks and landing places but largely different intensities and shapes. At landing time, TR reached a Category 6 with 60 m s^{-1} maximal winds and 910 hPa air pressure, while TK reached Category 4 with 40 m s^{-1} maximal winds and 960 hPa air pressure (Figure 4). They both caused severe damage to the China coasts. TR killed at least 62 people and caused 37000 houses collapses and an economic loss of about US$ 6.25 billion. TK killed at least eight people and caused 1800 houses collapse and an economic loss of about US$ 55.86 million. Figure 3 show house damages and surge at the Zhanjiang Coast during TR and TK, respectively.

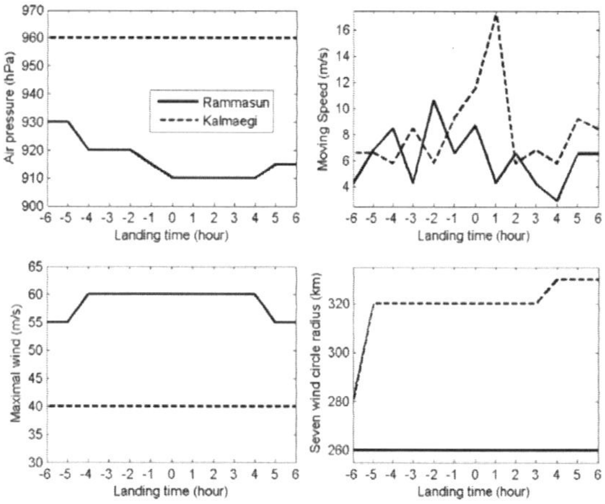

Figure 4. Central air pressure, moving speed, maximum wind, and seven wind circle radius during TR (solid lines) and TK (dashed lines) before and after landing time.

3. Storm Surge Comparison

TR and TK both generated storm surges along the Zhanjiang Coast when they landed at Wenchang. Figure 5 shows the instantaneous (hourly) storm surges, tides, wind speeds, and air pressures during the two typhoons at the Zhapo, Shuidong, Zhanjiang, and Naozhou tide gauge stations along the Zhanjiang Coast. It can be seen that TR caused the highest storm surge at Naozhou, where the peak of the storm surge reached 2.60 m. Surprisingly, the maximum storm surge of 4.50 m was caused by TK at the same tide gauge station, although the intensity of TK (Category 4) was much weaker than TR (Category 6). Higher peak surges also have been found at Zhanjiang (4.35 m), Shuidong (3.00 m), and Zhapo (2.20 m) during TK, compared with those during TR (2.56 m, 1.60 m and 1.15 m, respectively), indicating the height of the storm surges was not proportional to the typhoons' intensity in some cases along the Zhanjiang Coast. We provide specific analyses about why the weaker TK caused a much higher storm surge compared with the stronger TR in the discussion section.

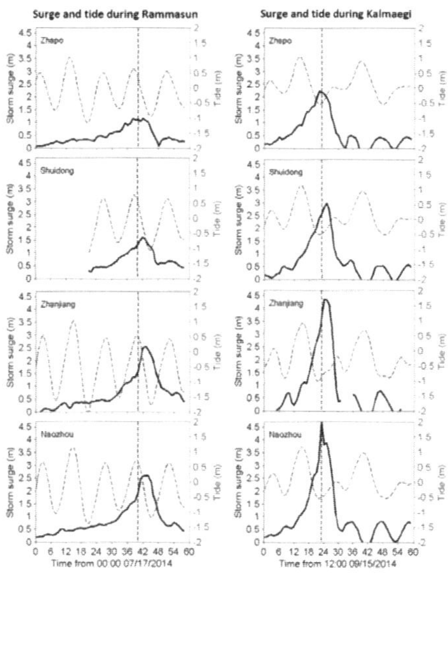

Figure 5. Storm surge (solid lines) and tide (dashed lines) at the Zhapo, Shuidong, Zhanjiang and Naozhou tide gauge stations along the Zhanjiang Coast during TR (left panel) and TK (right panel). Vertical dashed lines represent landing times.

4. Discussion

4.1. Atmospheric forcing

It is well known that height of a storm surge is proportional to the square of wind speed:

$$\eta \propto w^2/H, \qquad (1)$$

in which η and w are the surge height and wind speed, respectively (Powell and Reihold, 2007). Figures 6 shows that the peak wind speeds (minimum air pressures) of TK were much larger (lower) than that of TR at all tide gauge stations along the Zhanjiang Coast. In addition, all the tide gauge stations were located at the northwest of the typhoon centers before landing time, which made the wind direction similar. Figure 7 shows that wind direction switched from the northern direction to the northeastern direction at all tide gauge stations during TR and TK, which paralleled the shelf of the Zhanjiang Coast and was favorable for the Ekman transport of water mass toward the shore. These results indicate that the larger wind speeds and lower air pressures possibly contributed in some way to the higher surge during TK. The question of why atmospheric forcing during the lower category TK storm was stronger than that of the higher category TR storm at these tide gauge stations can be explained by the distribution of their maximum wind circle radii. Jelesnianski et al. (1992) purposed a formula for calculating the wind speed profile of a stationary storm, which is described as:

$$V_r = V_{max} \times (2 \times R_{max} \times r)/(R_{max}^2 + r^2), \tag{2}$$

in which V_{max} is the maximum wind speed, r is the radius, and $R_{max}$is the radius of maximum wind. The formula for the air pressure profile is the same as for the wind. The authors suggested that the wind speed and air pressure in the outer circle of a typhoon are not only determined by typhoon intensity, but also by the typhoon maximum wind radius. It can be seen that TR and TK had similar ten wind circle radii at between 140 and ~160 km (not shown). However, TK had a much larger seven wind circle radius (about 320 km) than that of TR (about 260 km) (Figure 4), which suggests the power of TR was weakened much faster than that of TK when the wind circle extended outward to these tide gauge stations along the Zhanjiang Coast. As a result, more severe atmospheric conditions occurred at this region during TK.

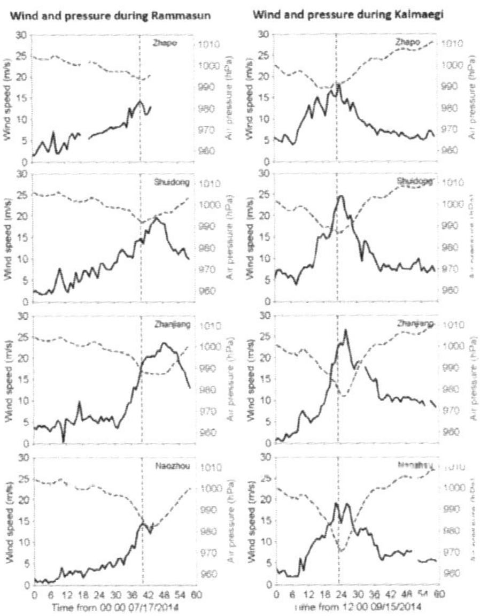

Figure 6. Wind speed and air pressure at the Zhapo, Shuidong, Zhanjiang, and Naozhou tide gauge stations along the Zhanjiang Coast during TK (solid lines) and TR (dashed lines). Vertical dashed lines represent landing times.

4.2. Typhoon moving speed

Proudman (1953) suggested storm surges reach maximum value when the moving speed is close to the propagation speed of the long wave, i.e.

$$\frac{1}{1-\frac{U^2}{gh}} \rightarrow \infty, \tag{3}$$

in which U is the typhoon moving speed and $\sqrt{gh}$ is the long wave propagation speed. The mean depth of the Zhanjiang Coast is about 25 m and its long wave propagation speed is about 15.8 m s^{-1}. Figure 4 shows that TR and TK had similar moving speeds (about 7 m s^{-1}) at 1 hour prior to landing. After that, TK suddenly accelerated and its moving speed increased to 15 m s^{-1} at 1 hour after landing at the Wenchang. This was very close to the long wave propagation speed. Conversely, TR decelerated very fast to about 4 m s^{-1} at 1 hour after landing at the Wenchang, a speed that was much smaller than the long wave propagation speed. As a result, a much higher storm surge was generated along the Zhanjiang Coast during TK compared with that of TR. This phenomenon also has been found at other worldwide coastal areas (Rego and Li, 2009; Zhang et al., 2012). Note that the storm surge was generated and dissipated much quicker during TK than TR. Figure 5 shows that the time for storm surge generation from zero to peak was about 24 hours and reduced from peak to zero in about 6 hours during TK. But these times were more than 42 hours and 12 hours for TR, respectively. These results suggest that typhoon with faster moving speed can fleetly cause higher storm surge, but the storm surge declines quicker after its passing through the Zhanjiang Coast.

4.3. Tide–surge interaction

Tide–surge interaction has been recognized for the past half century. Tide not only affects the storm surge height, but also changes the storm surge phase through linear and nonlinear interactions (Rossiter, 1961; Wolf, 1978; Jone and Davies, 1998; Shen et al., 2006; Rego and Li, 2010, Xu et al., 2010). Usually, flood tide produces higher storm surges compared with ebb tide, and the linear superposition of storm surge and tide can represent total water elevation during typhoons in coastal regions (Proudman, 1957). In this study, we separate the tide and storm surge to see how tide impacted the storm surge during TR and TK. Figure 5 shows that the storm surge reached its peak at the ebb period of spring tide during TR, and the declining tide retarded the development of the storm surge and reduced the total water level along the Zhanjiang Coast. Contrarily, storm surge reached its maximal value at the flood period of neap tide during TK, and the rising tide had a positive effect on development of the storm surge in this region. It is understood that the effect of nonlinear tide–surge interaction on the height of a storm surge is mainly due to alteration of bottom friction. Positive tide increases the total water level and reduces bottom friction, which is constructive to storm surge, while negative tide does the opposite (Davies et al. 2000; Xu et al., 2010). Figure 5 shows that tide level was higher than the mean water level, which increased total water depth and reduced bottom friction during TR. The tide level during TK was lower than the mean water level, which decreased total water depth and added bottom friction, suggesting that the nonlinear tide–surge interaction was destructive to the storm surge during this period. However, how much impact of the nonlinear tide–surge interaction had on storm surge during the two typhoons is unable to be estimated by currently observed results. To explore this, more studies using numerical models will be needed in the near future.

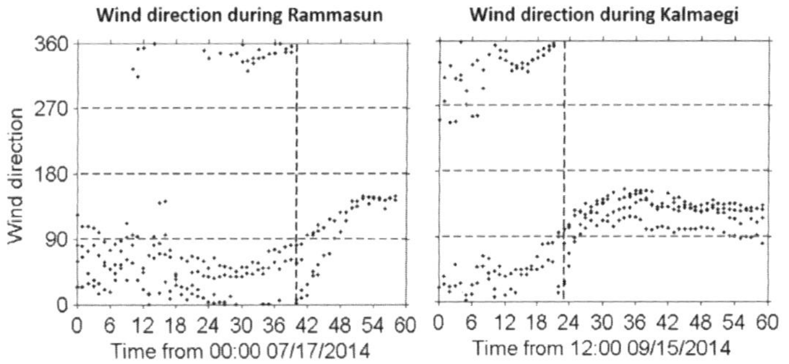

Figure 7. Wind directions (degree) at all four gauge stations during TR and TK. Direction 0, 90, 180 and 360 represent northern, eastern, southern and western winds. Vertical dashed lines represent landing times.

5. Summary

The distributions and propagation processes of storm surges along the Zhanjiang Coast during TR and TK are reported in this study. We find that the peak of the storm surge during TK was significantly larger than that of TR in this region, although the intensity of TK was much weaker than TR. The question of why a higher storm surge was generated during TK than TR can be answered by the three combination of three factors : 1) The power of TK was weakened when the maximum wind circle extended outward to the Zhanjiang Coast, causing stronger winds and lower air pressure during TK compared with TR in this region; 2) the moving speed of TK was closer to the long wave propagation speed compared with TR, which is a favorable factor for high storm surge; and, 3) TR passed by the Zhanjiang Coast during flood tide, which increased its storm surge, while TK passed by this region during ebb tide, which reduced its storm surge. Although the tide–surge nonlinearity had negative and positive effects on the storm surge during TK and TR, respectively, it may have played only a minor role during these processes. To understand speficic impacts of these factors on storm surge along the Zhanjiang Coast, numerical study will be applied in our future work.

Funding Statement

This work was funded by the Knowledge Innovation Program of the Chinese Academy of Sciences under Grants SIDSSE-201306 and SIDSSE-201205, the National Natural Science Foundation of China under Grants 41406005, and the "Hundred Talents Program" of the Chinese Academy of Sciences under Grant SIDSSE-BR-201304 and Y410161KYF.

References

[1] Cao, J.F., Wu, D.P., Lang, X.B., 2006. The storm surge analysis on Typhoon "Krovanh". Marine Forecasts. 23(2), 63-66.
[2] Chen, Y.D., Dong, Z.J., Jiang, G.R., Luo, J., 2002.Characteristic analysis of Zhanjiang Harbor's storm surge. Marine Forecasts. 19 (3), 44-52.

[3] Davies, A.M., Kwong, S.C.M., Flather, R.A., 2000.On determining the role of wind wave turbulence and grid resolution upon computed storm driven currents. Continental Shelf Research. 20(1), 825–1,888.
[4] Dong, J.X, Li, T., Hou, J.M., Yu, F.J., 2014. The characteristics of temporal and spatial distribution of storm surge in Guangdong Province and storm surge hazard study at Yangjiang City. Acta Oceanologica Sinica (in Chinese). 36(3), 83-93.
[5] Elliott, R.J.R., Strobl, E., Sun, P., 2015. The local impact of typhoons on economic activity in China: A view from outer space. Journal of Urban Economics. 88, 50 66.
[6] Han, S.Z., Pan, S., 2013. Numerical Simulation and Analysis of Storm Surge in the Hangzhou Bay. Periodical of Ocean University of China. 43(7), 1-6.
[7] Jelesnianski, C.P., Chen, J., Shaffer, W., 1992. SLOSH: Sea, Lake, and Overland Surges from Hurricane. National Weather Service, Silver Springs, MD, (71p).
[8] Jones, J.E., Davies, A.M., 1998. Storm surge computations for the Irish Sea using a three-dimensional numerical model including wave–current interaction. Continental Shelf Research. 18(2), 201–251.
[9] Peng, M.C., Xie, L.A., Pietrafesa, L.J., 2004. A numerical study of storm surge and inundation in the Croatan–Albemarle–Pamlico Estuary System. Estuarine, Coastal and Shelf Science. 59(1), 121-137.
[10] Powell, M.D., Reinhold, T.A., 2007.Tropical cyclone destructive potential by integrated kinetic energy. Bulletin of the American Meteorological Society. 88(4), 513-526.
[11] Proudman, J., 1957. Oscillations of tide and surge in an estuary of finite length. Journal of Fluid Mechanics. 2(4), 371–382.
[12] Proudman, J., 1953. Dynamical Oceanography. John Wiley, New York, 409.
[13] Rego, J.L., Li, C., 2009. On the importance of the forward speed of hurricanes in storm surge forecasting: A numerical study. Geophysical Research Letters. 36(7), doi:10.1029/2008GL036953.
[14] Rego, J.L., Li, C., 2010. Nonlinear terms in storm surge predictions: Effect of tide and shelf geometry with case study from Hurricane Rita. Journal of Geophysical Research. 115(C6), doi:10.1029/2009JC005285.
[15] Rossiter, J.R., 1961. Interaction between tide and surge in the Thames. Geophysical Journal International, 6(1), 29-53.
[16] Shen, J., Gong, W., Wang, H.V., 2006. Water level response to 1999 Hurricane Floyd in the Chesapeake Bay. Continental Shelf Research. 26(19), 2484–2502.
[17] Shi, J., Yang, Z., Sha, W., 2004. Discussion on relation between surge and such elements as landing location and path of tropical cyclone effected on Zhanjiang Harbor. Marine Forecasts (in Chinese). 21(1), 43-51.
[18] Weisberg, R.H., Zheng, L., 2006. Hurricane storm surge simulations for Tampa Bay. Estuaries and Coasts. 29(6), 899-913.
[19] Wolf, J., 1978. Interaction of tide and surge in a semi - infinite uniform channel, with application to surge propagation down the east coast of Britain. Appl. Math. Model. 2(4), 245–253.
[20] Xu, H.Z., Zhang, K.Q, Shen, J., Li, Y.P, 2010. Storm surge simulation along the US East and Gulf Coasts using a multi-scale numerical model approach. Ocean Dynamics. 60(6), 1597-1619.
[21] Zhang, K.Q, Liu, H.Q, Li, Y.P, Xu, H.Z, Shen, J., Rhome, J., Smith, T.J., 2012. The role of mangroves in attenuating storm surges. Estuarine, Coastal and Shelf Science. 102, 11-23.

Elastic and Electrical Properties Evaluation of Low Resistivity Pays in Malay Basin Clastics Reservoirs

Luluan Almanna Lubis[1,2], Deva P. Ghosh[1,2], Maman Hermana[2]
[1] Geosciences Department, Universiti Teknologi PETRONAS, Malaysia
[2] Centre of Excellence in Subsurface Seismic Imaging and Hydrocarbon Prediction, Universiti Teknologi PETRONAS, Malaysia

E-mail: luluan.lubis@petronas.com.my

Abstract. The elastic and electrical properties of low resistivity pays clastics reservoirs in Malay Basin are strongly dependent on the complex nature of the clay content, either dispersed or laminated/layered. Estimating the hydrocarbon pore volume from conventional electrical log, i.e. resistivity log, is quite a challenge. The low elastic impedance contrast also found as one of the challenge thus create a problem to map the distribution of the low resistivity reservoirs. In this paper, we evaluate the electrical properties and elastic rock properties to discriminate the pay from the adjacent cap rock or shale. Forward modeling of well log responses including electrical properties are applied to analyze the nature of the possible pays on laminated reservoir rocks. In the implementation of rock properties analysis, several conventional elastic properties are comparatively analyzed for the sensitivity and feasibility analysis on each elastic parameters. Finally, we discussed the advantages of each elastic parameters in detail. In addition, cross-plots of elastic and electrical properties attributes help us in the clear separation of anomalous zone and lithologic properties of sand and shale facies over conventional elastic parameter crossplots attributes. The possible relationship on electrical and elastic properties are discussed for further studies.

1. Introduction

For the last forty years, the Malay Basin has been a prolific setting for oil and natural gas explorations in south east asia which focused on several plays: structural play, stratigraphic play and basement plays (Ghosh et.al., 2010). In the Malaysian Basins, the economical important of low resistivity pay sands just recently demonstrated. Riepe et.al. (2012) and Kantaadmadja et.al. (2014) are few authors who reveal the important of these pays in the Malaysian Basins. These pays may revealed ten to hundreds million barrels of hydrocarbon (Kantaadmadja et.al., 2014). The similar problems of bypassed pays also mentioned as one of the geophysical challenges by the other authors (Ghosh et.al., 2010).

[1] Geosciences Department, Universiti Teknologi PETRONAS, Malaysia, Bandar Seri Iskandar, 32610, Tronoh, Perak Darul Ridzuan, Malaysia

The low resistivity pays are generally documented, due to the low resistivity and low contrast reading from the resistivity log. There are several factors associated to the low reading of the resistivity log fresh formation water or low salinity micro- porosity, conductive mineral as part of the minerals in the rock and very thin inter-bedding of sand and shale formation. (Worthington, 2000)

The formation parts for instance lithology, fluid and pore space of the rock influence by the depositional and geological setting, thus directly affect the responses of resistivity log and seismic data. Because of the complex nature of low resistivity pay triggering the geophysical experimental, interpretation and numerical studies turn out to be extremely challenging. Therefore this problem has mystified petrophysicist, geologist as well as geophysicist to solve the problems.

The solution of the problems requires integrated rock properties analysis, both electrical and elastic, to tackle complex mineralogy and poor understanding of anisotropic conditions to reexamine the current understanding of the reservoir properties of low resistivity pays. In this study an integrated rock properties analysis on electrical properties forward modeling and elastic rock properties analysis to discriminate the pay from the adjacent shale are tested through a rigorous workflow, started from well log interpretation, forward modeling of the well log responses to the elastic rock properties analysis.

2. Literature Review

2.1. Low Resistivity Pay Zone

Several numbers of technical papers described on the possibility of finding the low resistivity pays. These pays are possible to find in almost all major siliciclastic depositional environments (Kantaadmadja et.al., 2014). The most general causes of low resistivity pays are the clay minerals. The clays which filled by water in its microporosity can be acted to exchange cations within the pore fluids which caused the suppression of the electrical logs reading (Worthington, 2000).

The pay zones that produce low resistivity or low-contrast log responses are inclined by a variation of aspects related with anisotropy, conductive mineralogy, water salinity or fresh water formation in addition to microporosity, dip and bed thickness. There are several measurements can be done to resolved bed thickness effect. One of the technology are the use of borehole image logs and triaxial induction resistivity (Claverie et al., 2010). The summary of the log responses of the formations in the Malay Basin are well reported by Kuttan, M. (1980). Most of the cases related to the low resistivity pay are the reservoirs with high silt even though it is low clay content and very fine grained sand. The fresh water formation affect the log responses as well.

2.2. Electrical and Elastic Properties Relationship

In normal condition, the pay zone can be easily distinguish from the adjacent shale and the water-bearing zone by using resistivity log. The good illustration from http://www.spec2000.net is shown below:

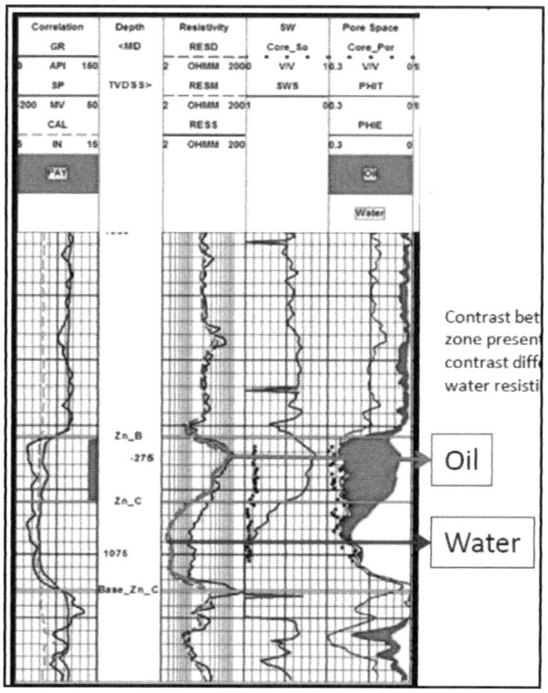

Figure 1: Contrast between oil and water zone presents because of the contrast difference between oil and water resistivity (from http://www.spec2000.net)

Contrast between oil and water zone presents because of the contrast difference between oil and water resistivity. The tool works by conducting electric current, and the readings are depending on the type of lithology and pore fluids.
Generally, the velocities depend on the ratio rock moduli to density. The moduli and density effects generate velocity variations which depend on the rock properties such as porosity, fluid saturation, and texture. The seismic (P- and S- wave) velocities – Vp and Vs respectively – of an isotropic rock material can be estimated using known rock moduli and density as follows:

$$V_P = \sqrt{\frac{K_{sat} + (4/3)\,\mu_{sat}}{\rho}},$$

$$V_S = \sqrt{\frac{\mu_{sat}}{\rho_{sat}}},$$

where K is bulk modulus, μ is shear modulus, and ρ is the bulk density.

The physics of elastic and electrical properties in conducting a porous media share no physical parameter. Thus this make the velocity and resistivity cannot be directly related (Carcione, 2007). Rock physics information is usually used to link the elastic and electrical properties data through porosity. This link is not exact (Werthmuller, 2014).

Some of the recent approaches are by developing joint inversion for CSEM and seismic data through rock physics and structural links (Abubakar, 2012). The step they have done is to first determine porosity and saturation of the formation. The results of the inversion was claimed to be better than a single dataset inversion, seismic or CSEM data. Werthmuller (2014) used background resistivity model which derived from velocities data. However the weaknesses concluded from the paper are the model is isotropic and having no resistivity calibration outside well logs data location. The approach by Abubakar, 2012, was adopted by Gao using crosswell synthetic data. Dell'Aversana, P., G (2012) used velocity, resistivity and density measurement from crosswell and involving bayes to link electrical and elastic properties. Kwon and Snieder (2011) investigate uncertainty in the combination of CSEM and seismic data. The approach is by using a synthetic and Bayesian. They consider several type of sources of uncertainty, from well log are seismic wave velocity and electrical resistivity. Seismic and CSEM data are the other data consider for the sources of uncertainty.

There are several researchers documented the relation between electrical and elastic properties. Soldal (2015), CO2 flooding experiment of reservoir sandstones-monitoring changes in acoustic and electric properties. Cheng, (2015), Building a Model of Coupled Elastic and Flow Properties of Porous Rock. Ong, O, et.al., (2015), Seismic anisotropy and elastic properties of a VTI medium. Tang, X, et.al., (2015), Assessing rock brittleness and fracability from radial variation of elastic wave velocities from borehole acoustic logging.

3. Methodology

3.1. Data set

The principal data sources for electrical forward modeling and elastic rock properties analysis are the well logs data and core information. Log data includes compressional and shear sonic logs, density, gamma ray, and resistivity. We incorporated the core and petrophysical study that has already been done by PETRONAS (Malaysia) which contains a series of petrophysical properties interpretation that are the key for calibrating and improving the electrical forward modeling and elastic rock properties analysis.

3.2. Quick-look Well log Interpretation

The objective of well log interpretation to locate and define a given reservoir and also quantitatively the thickness of the reservoir, porosity and water saturation.

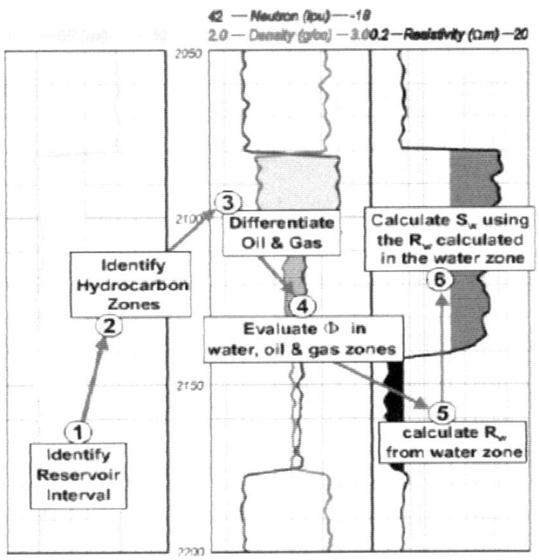

Figure 2: Steps to identify the pay zone using well log data

Figure 2. shows the steps to identify the pay zone. The first step is to identify the reservoir interval through GR log readings. Low GR usually represent sand, or for some cases coal and carbonates. Next step is to identify the hydrocarbon zones by interpreting the resistivity and neutron – density data. The crossover on neutron – density logs represent gas reservoirs with resistivity reading is high. The evaluation of the porosity of the reservoir can be calculated from neutron and density logs. The last step for quick-look interpretation is to calculate the water saturation of the pay zone.

3.3. Electrical Properties Forward Modeling

The forward modeling is constructed based on the specific petrophysical properties of the Malay Basin field example. The input for the forward modeling are water saturation (Sw), N/G, porosity, m and n value (cementation and saturation exponents respectively), GR cutoff for sand and resistivity of sands and shales end member. The electrical properties responses are simulated by applying a convolution filter to the beds conductivity.The neutron-density calculated from porosity input. The GR responses correspond to the beds created from the simulations.

3.4. Elastic Properties Cross-plot

The cross-plot of elastic properties can be used to interpret the lithology and pore fluid. Odegaard and Avseth (2003) introduced the rock physics templates (RPT) where the common elastic properties information used for RPT are velocity ratio which is compressional and shear velocity (Vp/Vs) ratios and acoustic impedance (AI). The used of different rock physics models combined with RPT also can be used for different type of lithology, such as carbonate (Lubis and Ghosh, 2014). Elastic properties cross-plot analysis is the first step to efficiently differentiate the lithology and pore fluids. The results of this analysis later on can be used as an input for AVO inversion. These elastic properties calculated and derived using the information from compressional and shear log data. The common quantitative analysis from the calculated elastic parameters (Vp/Vs and AI) indicated that the hydrocarbon pay sands, especially gas sands will have lower values compare to the adjacent shales and brine sands. The tabulated information for elastic properties cross-plots is shown on Table 1.

Table 1. Elastic properties and its derivation

Elastic Properties	Descriptions
Vp – Compressional Velocity	Compressional sonic log
Vs – Shear Velocity	Shear sonic log
AI – Acoustic Impedance	AI = ρ*Vp
SI – Shear Impedance	SI = ρ*Vs
Vp/Vs – Velocity ratio	Compresional sonic log/Shear sonic log
LambdaRho - Lame parameter*density	LambdaRho = $AI^2 - 2SI^2$
MuRho - Shear modulus* density	MuRho = SI^2
inverse Qp	$SQp^{-1} \equiv \frac{5}{6}\frac{1}{\rho}\frac{(M/G-2)^2}{(M/G-1)}$
inverse Qs	$SQs^{-1} \equiv \frac{10}{3}\frac{1}{\rho}\frac{(M/G)}{(3M/G-2)}$

4. Result and Discussion

The results obtained in this study can be divided into 3 parts, 1) quick-look well log interpretation, 2) electrical forward modeling and 3) Elastic properties cross-plot analysis. Figure 3 illustrates the basic well log interpretation from one of the well in Malay Basin. On Figure 2, Formation B 100 and D 35 are typical log responses of gas pay reservoirs. The main log responses to note is that over the sand interval at B100 and D 35, the Gamma Ray (GR) is low GR and having cross-over on the density and neutron responses. The resistivity showing high readings compare to the adjacent shale. Based on a standard log analysis the high GR on formation B 100 and D 35 would be interpreted as sand formation with high clay content. However, the interpretation with cautions should be considered since the work by Kuttan, M., (1980), reported the response can be from sand reservoirs with very fine grained and have high silt and low clay content. The elastic logs, AI and Vp/Vs showing low readings, typical gas sand reservoirs.

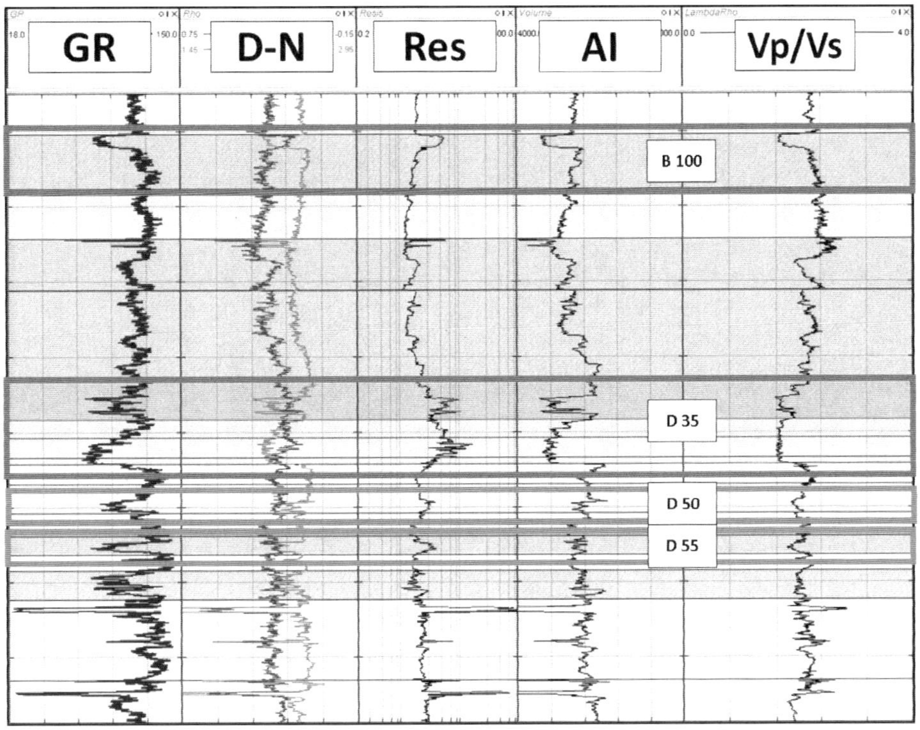

Figure 3. Quick-look interpretation on formation B and D in a field in Malay Basin.

Formation D 50 and D 55, are the oil bearing section (petronas report). This reservoir is atypical of low resistivity pay, where the resistitiy contrast with the adjacent shale are not having high contrast. The GR values show low readings for sand formation. The formation would be interpreted as shaly sandstone or another possibility as very fine grained sandstone with high silt and low clay content. The formation waters if it is fresh can cause the low contrast between the oil bearing reservoir and the adjacent shale.

Figure 4. shows forward modeling of electrical properties and other log responses for possible pay zones at B and D formation which interpreted from Figure 3. The yellow and green color in the track between GR and resistivity curves are the sands and shales, respectively. The blue color on GR track is showing the GR responses on each scenarios. The dashed red line on GR track is the cutoff used to determine the net to gross (N/G). The dashed line is halfway between clean-sand and shale end points. The result of net sands are shaded in tan yellow in GR tract. The electrical properties response, in these tracts on resistivity log responses, simulated by applying a convolution filter to the thin-bed conductivity values which is shown in Track 2. Track 3, simulated the responses of density and neutron logs. The forward modeling on Figure 4.a, simulated the log responses on formation B 100, where the GR, resistivity and density-neutron log responses are typical of gas sand reservoirs. The input for the forward modeling are 10% water saturation (Sw), 90% N/G, 30% porosity, m and n value are 2 (cementation and saturation exponents respectively), GR cutoof sand 60 GAPI, resistivity of sands and shales end member are 44 and 1 ohmm. The forward modeling results of the resistivity is around 13 ohmm.

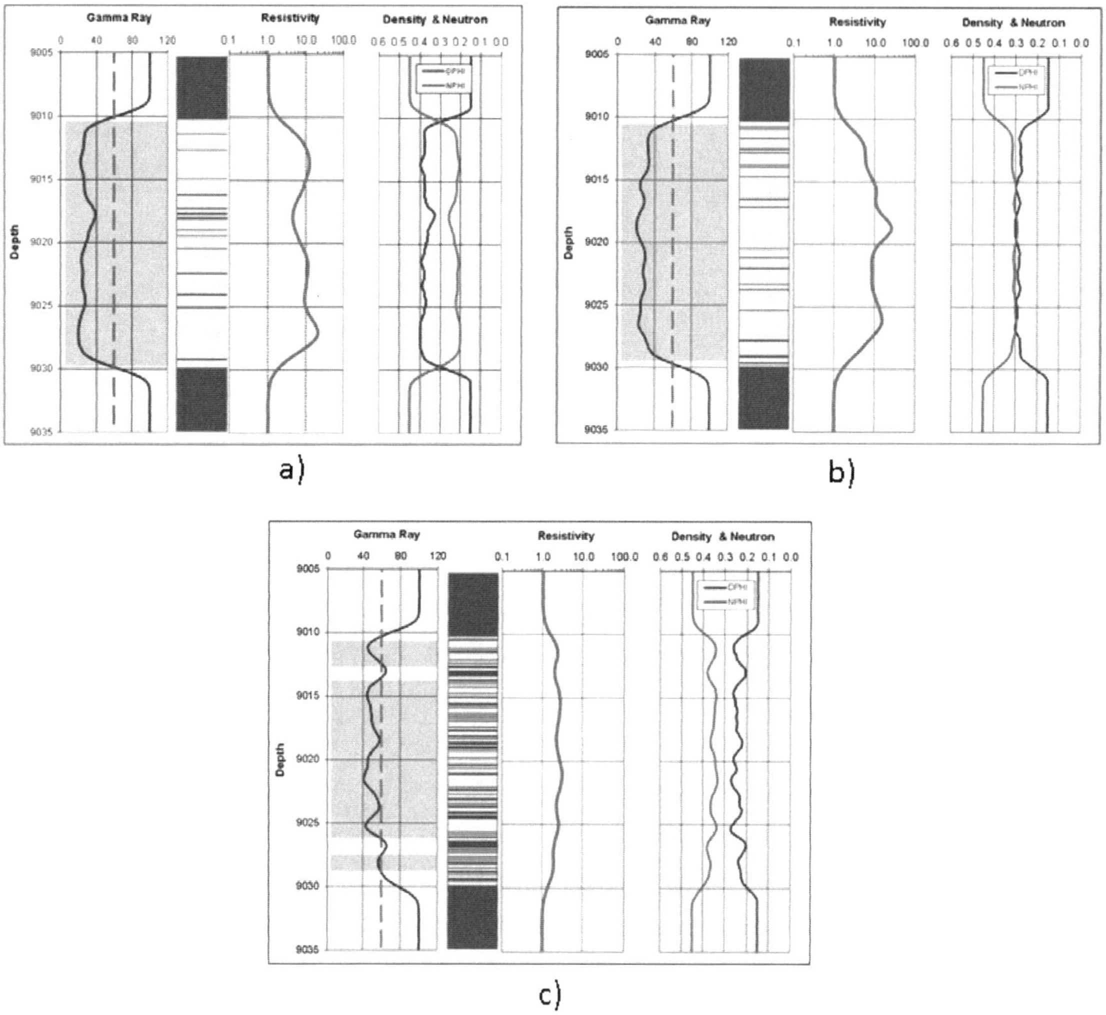

Figure 4. Forward modeling on well log responses, a) gas sand, b) oil sand and c) low resistivity pay.

The forward modeling on Figure 4.b, simulated the log responses if having a formation with oil sand pay. The input for the forward modeling are 90% Oil saturation (So), 90% N/G, 30% porosity, m and n value are 2 (cementation and saturation exponents respectively), GR cutoof sand 60 GAPI, resistivity of sands and shales end member are 44 and 1 ohmm. The forward modeling results of the resistivity is around 11 ohmm. The forward modeling on Figure 4.c, simulated the log responses on formation D50, for low resistivity pay with lower N/G. The input for the forward modeling are 10% water saturation (Sw), 60% N/G, 30% porosity, m and n value are 2 (cementation and saturation exponents respectively), GR cutoof sand 60 GAPI, resistivity of sands and shales end member are 44 and 1 ohmm. The forward modeling results of the resistivity is around 2 to 3 ohmm, typical low resistivity pay on the evaluated field.

The sands and shales are shown in yellow and green, respectively, in the track between the GR and resistivity curves. Track 1 shows the gamma ray in blue. The cutoff used to determine net sand is the red dashed line halfway between the clean-sand and shale endpoints. The resulting calculated net sand intervals are shaded tan. The electrical properties response, resistivity log, simulated by applying a convolution filter to the thin-bed conductivity values, is shown in Track 2; the simulated density and neutron logs are in Track 3.

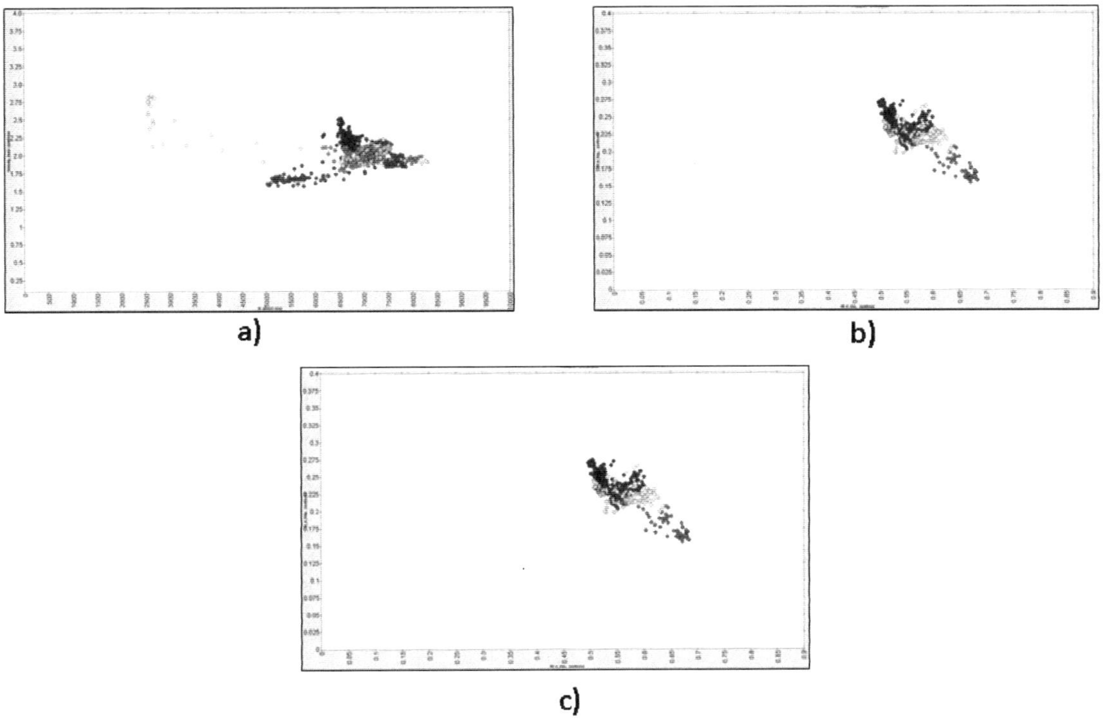

Figure 5. Elastic properties cross-plot analysis a) Vp/Vs vs AI, b)LambdaRho-MuRho and c)Sqp vs Sqs.

5. Conclusion

The integrated study on electrical and elastic properties analysis are presented and discussed in details. The objective of this study is to provide an interpretation of several pays in a field of Malay Basin, with focus on the low resistivity pay anomalies on elastic and electrical properties. Inverting the results directly to seismic data must be done with cautions due to the scale and dispersion effect. Thus this integrated analysis provides a preliminary study on transforming the elastic to electrical properties. The effect of different lithology and fluid types was separated from SQp domain and SQs domain respectively. The plots of lithology and fluid types effect are orthogonal on each domain axis. This approach can be used initially to model the resistivity from seismic velocity. The workflow can be started from the isotropic modeling the will be extended to anisotropic modeling for the velocity to resistivity transform. In our case the media is vertical transverse isotropy (VTI) for laminated sand shale sequences. The rock physics link can be used to develop the resistivity trend. SQs show potential inversion for resistivity modeling since its having similar responses with resistivity.

6. References

[1] Ghosh, D. et.al. "Geophysical issues and challenges in Malay and adjacent basins from an E&P perspective", The Leading Edge, April, 2010.

[2] L. Riepe, M. N. Zain and N. S. Zainudin., "An Integrated Petrophysical Analysis of Low Resistivity Low Contrast (LRLC) Pays in Clastic Reservoirs in Se Asia", PGCE, Kuala Lumpur, April, 2012.

[3] Kantaatmadja, B.P. et.al., "Unlock Hydrocarbon Volumetric Potential of LRLC Clastic Reservoirs in Malaysian Basins", IPTC, 2014.

[4] Worthington, P.F. "Recognition and evaluation of low-resistivity pay", Petroleum Geoscience, Vol. 6, 2000.

[5] Claverie, M., et.al., "A New Look at Low-Resistivity and Low-Contrast (LRLC) Pay in Clastic Reservoirs", SPE, 2010.

[6] Kuttan, M. et.al. "Log Interpretation In The Malay Basin", SPWLA, July, 1980.

[7] http://www.spec2000.net

[8] Carcione, J. M., B. Ursin, and J. I. Nordskag, 2007, Cross- property relations between electrical conductivity and the seismic velocity of rocks: Geophysics, 72, no. 5, E193–E204,.

[9] Werthmüller, D., A. Ziolkowski, and D. Wright, 2013, Background resistivity model from seismic velocities: Geophysics, 78, no. 4, E213–E223

[10]Abubakar, A., G. Gao, T. M. Habashy, and J. Liu, 2012, Joint inversion approaches for geophysical electromagnetic and elastic full-waveform data: Inverse Problems.

[11]Dell'Aversana, P., G. Bernasconi, F. Miotti, and D. Rovetta, 2011, Joint inversion of rock properties from sonic, re- sistivity and density well-log measurements: Geophysi- cal Prospecting, 59, 1144–1154

[12]Soldal, M. et.al., "CO2 flooding experiment of reservoir sandstones-monitoring changes in acoustic and electric propertie", 3IWRP, Perth, 2015.

[13]Cheng, A., et.al.,"Building a Model of Coupled Elastic and Flow Properties of Porous Rock", 3IWRP, Perth, 2015.

[14]Ong, O. et.al., "Seismic anisotropy and elastic properties of a VTI medium", 3IWRP, Perth, 2015.

[15]Tang, X., et.al.," Assessing rock brittleness and fracability from radial variation of elastic wave velocities from borehole acoustic logging", 3IWRP, Perth, 2015.

[16] Ødegaard, E., and P. Avseth, 2004, "Well log and seismic data analysis using rock physics templates": First Break, 22, 37–43

[17] Lubis, L. A., Ghosh, D. P., Harith, Z. Z. T., 2014, "Carbonate Pore Type Classification by using Digital Rock Images with a Rock Physics Model, Offshore Sarawak", EAGE/FESM Joint Regional Conference Petrophysics Meets Geoscience

Market Motivations for Voluntary Carbon Disclosure in Real Estate Industry

KALU Joseph Ufere*, BUANG Alias and ALIAGHA Godwin Uche.
Department of Real Estate, Faculty of Geoinformation and Real Estate,
Universiti Teknologi Malaysia, 81310 UTM, Skudai, Johor Malaysia,
khalujosef@gmail.com

Abstract

Climate change mitigation in developing economies is a balancing act, between economic development and environmental sustainability. The need for market friendly determinants for low carbon economy, without compromising economic development is of essence. The aim of the study is to determine market friendly factors, which motivates voluntary carbon information disclosure, in the real estate industry. The study modeled economic factor with three variables and financial market factor with three variables against voluntary carbon information disclosure in the real estate industry. Structural equation modeling was used for the modeling and content analysis was used to collect data on the level of voluntary carbon information disclosure, from 2013 annual reports of 126 real estate sector companies listed in the Kuala Lumpur Stock Exchange (KLSE). The model achieved a good fit, and was acceptable prediction. The results show that financial market factor has a significant predictive influence on voluntary carbon disclosure. The application of the result is that financial market factor is has a significantly positive influence on companies' willingness to make voluntary carbon disclosure in the real estate industry. The result may be limited to the real estate industry that is highly leveraged on syndicated fund.

Keywords: Climate change mitigation, Real Estate, Carbon disclosure, Market Motivations, Malaysia

1. INTRODUCTION

The risk associated with climate change is been experienced in most countries and the need to measure and manage its effect is of prime importance. There is a strong association between the emission of carbon dioxide (CO2) and other greenhouse gases (GHG) and the threat of climate change (Freedman & Jaggi, 2005). The effect of climate change is evident with the rise in surface temperature and sea levels, heavier rains with floods, hurricanes and storms, in some cases heat waves and drought (Revi, 2008). There is a quest by companies to stay competitive by improving their environmental records through energy efficiency, waste reduction, and recycling. This is to better satisfy the needs and expectation of their customers, also to adjust to the business landscape in view of climate change, gain support from society, improve media coverage or a blend of these benefits. The built environment being different or alternate to the natural environment in that it is a manmade environment at the expense of the natural environment; comprising of commercial, residential, industrial and institutional estates as well as roads, sidewalks, green spaces, and connecting transit systems (Younger, Morrow-Almeida, Vindigni, & Dannenberg, 2008). Property developers, management and investment companies are at the forefront of the quest from mitigation of the effect of climate change, and want to adopt green projects in order to appeal to the needs and expectations of customers especially those with concerns toward the environment.

The world's carbon emission for energy use from properties was 3% per annum in 1999 to 2004. But recent reports show that carbon emission for energy use in commercial and residential properties is estimated at 71% (Younger et al., 2008). In the US about half of the total carbon emission is from industrial, commercial and residential properties. Property development increases carbon emission in other ways, besides alternating for the natural environment; through the property use, building orientation, land location, building plan. The building influences each other and the surrounding landscape, the construction materials and building design elements of the internal and external walls, coupled with the use of energy and water supply by the occupants of the property. The consumption of electricity supply within the property affects the environment and through other interactions between the property and the environment it influences the environment negatively. Carbon emission from properties are primarily for electricity, used for lighting, cooling, heating and other uses in a building, which increases demand for energy which is derived from fossil fuel which emit carbon dioxide (Brown, Southworth, & Stovall, 2005). Real estate properties are long term structures so its emission of carbon is often for a long term. New buildings and older ones, have opportunity to integrate carbon reduction programs through energy efficiency for an eco-friendly neighborhood and environment. The cost for energy efficient project may involve high initial cost but offers long term savings and better ecosystem. The high initial cost could be an obstacle on the benefits of sustainable projects, but sustainable building development designs are cost effective and promote energy efficiency, protects the environment, and mitigate carbon emission that contributes to climate change (Younger et al., 2008). In developing economies the need to sustain economic development is weighed against the need for sustainable development. That is the reason for most countries are not willing to legislate for mandatory carbon disclosure and management which may have adverse effect on business and economic development.

There has being studies on the determinants for carbon information disclosure (Kansal, Joshi, & Batra, 2014; Luo, Lan, & Tang, 2012; Yunus, Elijido-Ten, Abhayawansa, & Vesty, 2016) with varying results. While Luo et al. (2012) argued that economic factors are deterministic, Kansal et al. (2014) claimed that among India companies social factors motivated companies decision for disclosure, but Yunus et al. (2016) held that the firm size and financial leverage are predictive factors. While Luo, Tang, and Lan (2013) found that there is a significant disparity between developed and developing countries as it concerns carbon disclosure, but most studies on the factor determinants for carbon disclosure are located in the developed countries and Amran, Periasamy, and Zulkafli (2014) in a study in Asia Pacific countries noted the need for an understanding on the determinants for voluntary carbon disclosure among the emerging economies due to the low level of disclosure noticed Therefore this study is to investigate market determinants that motivate companies in the real estate industry of a developing country.

2. LITERATURE REVIEW

Commercial sustainable property development is of a great advantage (Dixon, McNamara, & Newell, 2008) while the concept of sustainability has being around in the industry for some time, the implementation of the practice in property development, management and investment in most countries is still minimal (Sayce, Ellison, & Parnell, 2007). This is evident in the certificate for energy performance in the EU by property investors (Keeping, Dixon, & Ellison, 2007; Sturge, 2007). The property industry is observed by tenant as not making sufficient progress in relation to climate change mitigation programs (Bould, 2007). The major setback for sustainable real estate projects is the economic justification for the cost in commercial projects (Sayce et al., 2007), without established results on the relationship between sustainability and economic returns, it will be difficult to proceed with such projects.

The determinants for sustainable commercial building has being either through statutory legislation for environmental regulation, energy efficiency and carbon emission reduction from government in most developed countries or market factors (Keeping et al., 2007; Sayce et al., 2007; Sturge, 2007). The importance of the market factor is in its influence through the market mechanism influencing stakeholders at all levels, investor, business, government, tenant and community.

The Agency theory claims that financial market pressure comes from shareholders and debt holders to whom management is directly accountable. Failure to disclosure relevant information could lead to a wider information asymmetry between management and capital providers; thereby increasing the company's cost of capital (Cormier, Magnan, & Van Velthoven, 2005). Capital raising companies tend to provide more information on risk activities. The higher the leverage, the more pressure management would bear to disclose carbon information, it can be argued that companies that is highly leveraged tends to withhold some sensitive information, if the disclosure would increase transparency such that firm's financial risks would be higher than anticipated, thereby undermining the firm's negotiating position.

3. RESEARCH METHODOLOGY

Structural equation modeling (SEM) involves a mix of the confirmatory factor analysis and path models. It describes the measurement and structural models; evaluating the confirmation of hypothetical model and the interpretation of the final model (Chen, Zhang, Liu, & Mo, 2011). SEM permits the calculation of interrelated dependence and multiple relations, with the capacity to show unobserved models in these interactions and measurement error (Hair, 2010; Hair, Black, Babin, Anderson, & Tatham, 2009). The SEM is used in the study due to it can be used to develop a model and it robust multivariate results.

3.1 Sample selection

The data for the studies were drawn from the 2013 annual reports of property sector companies listed in Kuala Lumpur Stock Exchange (KLSE). The data was collected through content analysis on the level of disclosure; and financial ratios were used as constructs for the independent variables. Three industry-types selected within the sector were Property companies, Real Estate Investment companies, and the Construction companies. A total of 126 annual reports were selected, comprising sixteen (16) property investment companies, seventy six (76) property companies and thirty-four (34) construction companies. To measure the theoretical model the data was analyzed using Square-Structural Equation Modeling (SEM) instrument which is a more robust analysis tool than the regression model used in similar studies. Validity and reliability measures were applied. With the aim of theory development, SEM assesses the path relationships (Hair, Sarstedt, Hopkins, & Kuppelwieser, 2014).

3.2 Measures

The use of content analysis for the measurement of carbon disclosure in annual reports was adopted in Clarkson, Li, Richardson, and Vasvari (2008), (Freedman and Jaggi (2011)), and

Dwyer et al. (2009). Financial ratios were used to measure each factor social factor, institutional factor, economic factor, and financial market factor.

4. RESULTS

Measurement and structural models were employed, with the reliability and validity test applied to check before the structural model was examined. The analysis employed maximum likelihood for the estimate method as it provides a dependable approach to parameter assessment for a multiplicity of estimate state (Suki, 2014).

4.1 Structural model

Table 1: Model Fit Results

	χ^2	df	$\chi2/df$	GFI	AGFI	RMSEA	CFI	TLI
Recommended values			<3.0	>0.90	>0.90	<0.080	>0.90	>0.90
Model values	33.1144	24	1.3798	0.9474	0.9014	0.0551	0.9881	0.9822

The structural model in the SEM was performed and evaluated. A variety of indices were used to assess the model's overall fit (Table 1). The results indicate that the $\chi2$ of the model was 33.114 with 24 degree of freedom. The parsimonious fit category of the Chi Square/Degrees of Freedom ($\chi2/df = 1.3798$) was achieved, while the standard is to be less than 3; for the incremental fit category the Comparative Fit Index (CFI) of 0.9881 was achieved which requires a minimum requirement that is greater than 0.90. The Adjusted Goodness of Fit Index (AGFI) of 0.9014 was achieved against the threshold of greater than 0.90; for absolute fit, the Root Mean Square of Error Approximation (RMSEA) of 0.0551 with a target of less than 0.08, the Goodness of Fit Index (GFI) required a standard of greater than 0.900, and 0.9474 was achieved indicating a satisfactory model fit (Bentler, 1990; Byrne, 2001). Therefore the hypothesized model is of good fit, hence acceptable. So the results of the model are adequate and can be used for further perdition.

Table 2: Regression Weights

Path			Estimate	S.E	C.R	P
Financial Market Factor	→	Voluntary carbon disclosure	.1864	.8472	2.0336	.0420
Economic Factor	→	Voluntary carbon disclosure	.1493	11.6758	1.6241	.1044

$*p < 0.05$

5. DISCUSSION/CONCLUSION

The results show that the probability of getting a critical ratio as large as 2.0336 in absolute value is 0.0420. This means, the regression weight for financial-market factor in the prediction of Voluntary Carbon disclosure is significantly different from zero at the 0.05 level (two-tailed), hence has significant positive direct effect. While the probability of getting a critical ratio as large as 1.6241 in absolute value is 0.1044 for the economic factor. The implication of this result is that, the regression weight of Economic factors in the prediction of voluntary carbon disclosure in the real estate industry is not significantly different from zero at the 0.05 level, so no significant direct effect.

This means between the two markets factors modeled against voluntary carbon information disclosure financial market factor has a significant effect while the economic factor does not have, this result negates the finding in (Luo et al., 2012). Furthermore it means the variables within the financial market factor can be controlled to motivate market driven voluntary carbon information disclosure in the real estate industry, this agrees with (Yunus et al., 2016), towards low carbon economy and corporate climate change mitigation participation.

The implication of the results is that policy incentives can be developed around the financial market factor constructs to motivate companies' willingness to make voluntary carbon disclosure toward climate change mitigation.

This means that investors, capital market operators and regulator with the financial and banks can motivate climate change mitigation in the real estate sector by making mitigation requirements conditions for mortgage financing and for long term and bridging loan terms. This application is due to the significant positive direct effect between financial market factor and voluntary disclosure. This means as pressure increases from the financial market creditors for the real estate sector there will be a corresponding increase in voluntary carbon information disclosure, which involves measurement, monitoring and management of emission, this implications agrees with the postulations of agency theory.

References

Amran, A., Periasamy, V., & Zulkafli, A. H. (2014). Determinants of climate change disclosure by developed and emerging countries in Asia Pacific. *Sustainable Development, 22*(3), 188-204.

Bentler, P. M. (1990). Comparative fit indexes in structural models. *Psychological bulletin, 107*(2), 238.

Bould, R. (2007). New index for a new era. *Investment Property Focus, 7*, 7-9.

Brown, M. A., Southworth, F., & Stovall, T. K. (2005). *Towards a climate-friendly built environment*: Pew Center on Global Climate Change Arlington, VA.

Byrne, B. M. (2001). Structural equation modeling with AMOS, EQS, and LISREL: Comparative approaches to testing for the factorial validity of a measuring instrument. *International journal of testing, 1*(1), 55-86.

Chen, Y. Q., Zhang, Y. B., Liu, J. Y., & Mo, P. (2011). Interrelationships among critical success factors of construction projects based on the structural equation model. *Journal of Management in Engineering, 28*(3), 243-251.

Clarkson, P. M., Li, Y., Richardson, G. D., & Vasvari, F. P. (2008). Revisiting the relation between environmental performance and environmental disclosure: An empirical analysis. *Accounting, organizations and society, 33*(4), 303-327.

Cormier, D., Magnan, M., & Van Velthoven, B. (2005). Environmental disclosure quality in large German companies: economic incentives, public pressures or institutional conditions? *European accounting review, 14*(1), 3-39.

Dixon, T., McNamara, P., & Newell, G. (2008). The strategic significance of environmental sustainability by Australian-listed property trusts. *Journal of Property Investment & Finance, 26*(6), 522-540.

Dwyer, R., Lamond, D., Prado-Lorenzo, J.-M., Rodríguez-Domínguez, L., Gallego-Álvarez, I., & García-Sánchez, I.-M. (2009). Factors influencing the disclosure of greenhouse gas emissions in companies world-wide. *Management Decision, 47*(7), 1133-1157.

Freedman, M., & Jaggi, B. (2005). Global warming, commitment to the Kyoto protocol, and accounting disclosures by the largest global public firms from polluting industries. *The International Journal of Accounting, 40*(3), 215-232.

Freedman, M., & Jaggi, B. (2011). Global warming disclosures: impact of Kyoto protocol across countries. *Journal of International Financial Management & Accounting, 22*(1), 46-90.

Hair. (2010). Multivariate data analysis.

Hair, Black, W. C., Babin, B. J., Anderson, R. E., & Tatham, R. L. (2009). *Análise multivariada de dados*: Bookman Editora.

Hair, Sarstedt, Hopkins, & Kuppelwieser, V. (2014). Partial least squares structural equation modeling (PLS-SEM) An emerging tool in business research. *European Business Review, 26*(2), 106-121.

Kansal, M., Joshi, M., & Batra, G. S. (2014). Determinants of corporate social responsibility disclosures: Evidence from India. *Advances in Accounting, 30*(1), 217-229.

Keeping, M., Dixon, T., & Ellison, L. (2007). The energy performance of buildings directive and commercial property investment. *IPF, London.*

Luo, L., Lan, Y. C., & Tang, Q. (2012). Corporate incentives to disclose carbon information: Evidence from the CDP Global 500 report. *Journal of International Financial Management & Accounting, 23*(2), 93-120.

Luo, L., Tang, Q., & Lan, Y.-C. (2013). Comparison of propensity for carbon disclosure between developing and developed countries: A resource constraint perspective. *Accounting Research Journal, 26*(1), 6-34.

Revi, A. (2008). Climate change risk: an adaptation and mitigation agenda for Indian cities. *Environment and Urbanization, 20*(1), 207-229.

Sayce, S., Ellison, L., & Parnell, P. (2007). Understanding investment drivers for UK sustainable property. *Building Research & Information, 35*(6), 629-643.

Sturge, K. (2007). European property sustainability matters. *King Sturge, London.*

Suki, N. M. (2014). Passenger satisfaction with airline service quality in Malaysia: A structural equation modeling approach. *Research in Transportation Business & Management, 10*, 26-32.

Younger, M., Morrow-Almeida, H. R., Vindigni, S. M., & Dannenberg, A. L. (2008). The built environment, climate change, and health: opportunities for co-benefits. *American journal of preventive medicine, 35*(5), 517-526.

Yunus, S., Elijido-Ten, E., Abhayawansa, S., & Vesty, G. (2016). Determinants of carbon management strategy adoption: evidence from Australia's top 200 publicly listed firms. *Managerial Auditing Journal, 31*(2).

IOP Conf. Series: Earth and Environmental Science **38** (2016) 012006 doi:10.1088/1755-1315/38/1/012006

Modelling of Seismic and Resistivity Responses during the Injection of CO2 in Sandstone Reservoir

Muhamad Nizarul Idhafi Bin Omar [1], Luluan Almanna Lubis [1], Muhammad Nur Arif Zanuri[1], Deva P. Ghosh[1], Sonny Irawan[2], Shiferaw Regassa Jufar[2]
Geoscience Department, Universiti Teknologi PETRONAS, 31750 Tronoh, Perak, Malaysia.
E-mail: nizarulidhafi@gmail.com
Abstract: Enhanced oil recovery plays vital role in production phase in a producing oil field. Initially, in many cases hydrocarbon will naturally flow to the well as respect to the reservoir pressure. But over time, hydrocarbon flow to the well will decrease as the pressure decrease and require recovery method so called enhanced oil recovery (EOR) to recover the hydrocarbon flow. Generally, EOR works by injecting substances, such as carbon dioxide (CO2) to form a pressure difference to establish a constant productive flow of hydrocarbon to production well. Monitoring CO2 performance is crucial in ensuring the right trajectory and pressure differences are established to make sure the technique works in recovering hydrocarbon flow. In this paper, we work on computer simulation method in monitoring CO2 performance by seismic and resistivity model, enabling geoscientists and reservoir engineers to monitor production behaviour as respect to CO2 injection.

1. Introduction

Greenhouse gases has become an attention as the main cause of global warming. Due to global warming, reduction technologies for CO2, the most abundant greenhouse gas, become an urgent business. Kyoto Protocol (1997) defined an implementation goal that most of the world's countries could reduce CO2 emission by 5.2% of the 1990 totals. Prior to this, underground storage of carbon dioxide was introduced to reduce the effect of greenhouse gases to help in slowing down global warming [1].

The abundant of greenhouse gases often linked with fossil energy sources exploration, production and usage [2]. The carbon capture and storage (CCS) project, targeting on reducing CO2 in the atmosphere works by storing captured CO2 into geologic structure, such as large sedimentary basins in Illinois, Michigan and Western Canadian sedimentary basin [3]. In oil and gas exploration, the CO2 instead of being stored in large basins, also been widely used in enhanced oil recovery (EOR) to retrieve the recovered oil by performing a pressure difference to enable hydrocarbon migration to producing wells.

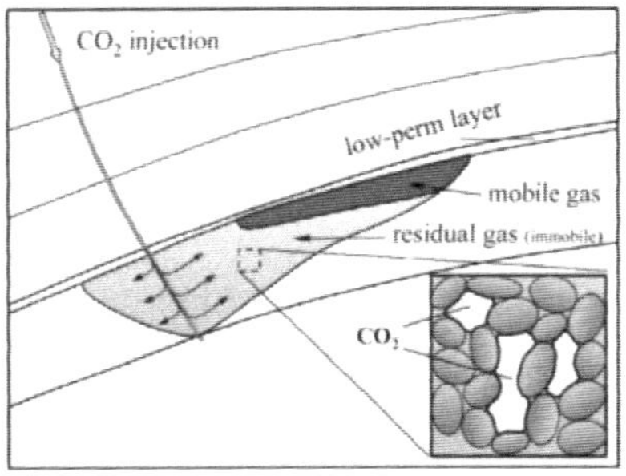

Figure 1.1: CO2 stored in saline aquifer for CCS project [4]

[1] Geosciences Department, Universiti Teknologi PETRONAS, Malaysia, Bandar Seri Iskandar, 32610, Tronoh, Perak Darul Ridzuan, Malaysia
[2] Petroleum Engineering Department, Universiti Teknologi PETRONAS, Malaysia, Bandar Seri Iskandar, 32610, Tronoh, Perak Darul Ridzuan, Malaysia

The method, by using CO2 in many cases is the most cost effective method to extract the final amounts of recoverable oil from depleted field [5] which share the same technology as carbon capture and storage (CCS) in term of development and monitoring [3]. Despite being conceptually the same, when implemented in EOR, the seismic and resistivity respond of the CO2 will not be the same as in CCS, where the CO2 now will be in contact with hydrocarbon to lead the recovery effort.

1.1 Enhanced Oil Recovery (EOR)

For over 40 years, injecting CO2 to tertiary reservoirs has been widely accepted as an effective method in enhanced oil recovery (EOR) [6]. Since October 1996, Statoil and its Sleipner partners working on Utsira Sand, offshore Norway have injected CO2 about 200 meter below reservoir top [1]. In Gullfaks Field in the North Sea, by injecting CO2 instead of water for EOR, significant amount of extra oil can be retrieved [7]. By using CO2 as an alternative for EOR, the majority of previous and current CO2 EOR projects has high economic return in high gas utilization efficiencies (167-227 cm^2 CO2/STB oil) [6].

In environmental perspective, the potential of CO2 storage combining with EOR is high, up to 60% of injected CO2 can be retained in the reservoir at the CO2 breakthrough if reinjection is not considered [6] . The criteria for suitable reservoir selection for CO2 EOR has been developed in [6].

Table 1.1: Optimum reservoir parameters and weighting factors for ranking oil reservoirs suitable for CO2 EOR [6].

Reservoir parameters	Optimum values	Parametric weight
API Gravity (°API)	37	0.24
Remaining oil saturation	60%	0.20
Pressure over MMP (MPa)	1.4	0.19
Temperature (°C)	71	0.14
Net oil thickness (m)	15	0.11
Permeability (mD)	300	0.07
Reservoir dip	20	0.03
Porosity	20%	0.02

1.2 Resistivity responds of CO2 injected sandstone

Electrical log has been widely used to provide quantitative correlation in exploitation of oil and gas reservoirs to provide an indication on the reservoir content [8]. Direct measurement of geophysical data during CO2 injection are not applicable due to technical and cost issues [9]. Therefore, indirect measurement consist of time lapse survey, resistivity as well as gravity were applied later on. The resistivity value is based on Archie's Law [8] generally will increase as CO2 concentration increase. This is due to resistance built up when more CO2 is added causing harder for the current to flow through the reservoir. The increment trend can be seen following the experiment in [10] [11] [12].

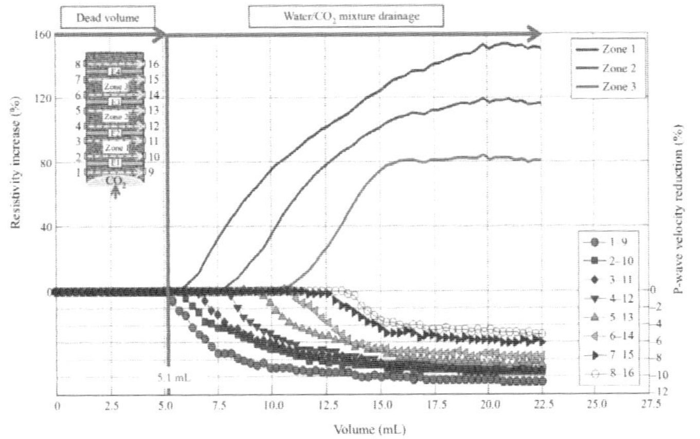

Figure 1.2: Changes in resistivity and P-wave velocities for three zones after CO2 injection. Note the increasing trend of resistivity as CO2 volume increase [10].

1.3 Seismic monitoring on CO2 injection in EOR

4D seismic exploration or time lapse reservoir monitoring (eg: Figure 1.3) has advanced rapidly over the past 10 years to monitor the progress of the injected fluid fronts (eg: water, gas, CO2, etc.) that can save hundreds of millions of dollars in optimizing injection programs [13]. A pilot project in Nagaoka was conducted and revealed after the injection of CO2, the maximum velocity experience 3.0% reduction after injecting 3200 t of CO2 and 3.5 % after injecting 6200 t of CO2 [14]. Notice that the increment of volume of CO2 injected were increased about two times, but the velocity reduction do not increase significantly.

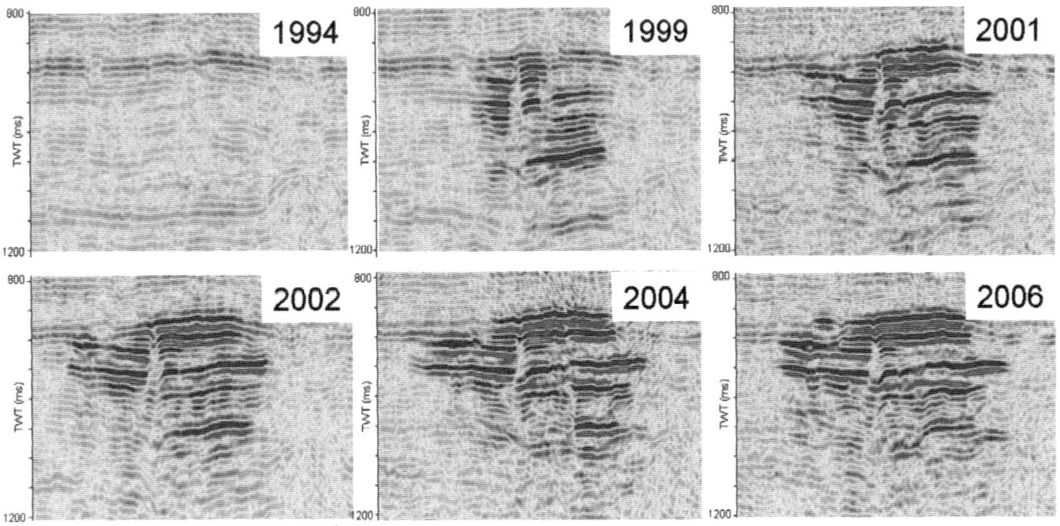

Figure 1.3: The development of CO2 plume over time, studied using time lapse seismic survey [1]

1.4 Elastic Bound

To have a more rational velocity-porosity relations, the gap between well logs and cores with geologic model, rock physics diagnostic analysis is used [15]. Expressed based on Hooke's Law [16], the elastic behavior of rock and the pore fluid become the fundamental of rock physics. Gassmann's fluid substitution theorem widely used to relates rock with fluid under three major assumptions: the rock is homogeneous and isotropic, the pore space is completely connected and the fluids are movable [17].

While many models proposed include complicated inputs, Gassmann relations reflect the changes in P and S wave velocity as accordance to saturation changes with simpler inputs. Density and P wave velocity logs are initial inputs and the imported parameter during substitution are matrix bulk modulus (Kmatrix), frame or dry rock bulk modulus (Krock), porosity (ø) and rock shear modulus (G) [18].

$$\frac{K_{\text{sat}}^{(2)}}{K_{\text{mineral}} - K_{\text{sat}}^{(2)}} - \frac{K_{\text{fluid}}^{(2)}}{\phi(K_{\text{mineral}} - K_{\text{fluid}}^{(2)})} = \frac{K_{\text{sat}}^{(1)}}{K_{\text{mineral}} - K_{\text{sat}}^{(1)}} - \frac{K_{\text{fluid}}^{(1)}}{\phi(K_{\text{mineral}} - K_{\text{fluid}}^{(1)})}$$

Fluid substitution formula by Gassmann [17] (1.1).

Since the shear modulus is insensitive to fluid changes, it will remain constant during the substitution [18].

Table 1.2: Elastic and electromagnetic properties of some elements [19].

Medium	K (GPa)	μ (GPa)	ρ (g/cm^3)	R (μm)	η (Pa s)	σ (S/m)
Clay	25	20	2.65	1	-	0.2
Sand grains	39	40	2.65	50	-	0.01
Brine	2.25	0	1.03	-	0.0012	12
CO_2	0.025	0	0.5	-	0.00002	0

$\phi = 25\%$, $r_1 = 10$ cm.

In order to substitute fluids it is necessary to establish their properties. In this case, the two fluids in discussion are water (or brine) and CO2. Density and bulk modulus values (eg: Table 1.2) could be obtained from laboratory measurements but in most of the cases some empirical relations are used such as the equations of Batzle and Wang [20]. These yield the density and bulk modulus knowing temperature, pressure, gravity of the fluid (°API in oil) and concentration (ppm in brine). With these input values, a good approximation of fluid properties can be made at certain conditions. Some researchers caution that using the Batzle and Wang [20] equations as well Gassmann relations (1.1) [17] which were developed for a gas-oil-water scenario, might not represent the reality of a CO2 fluid substitution.

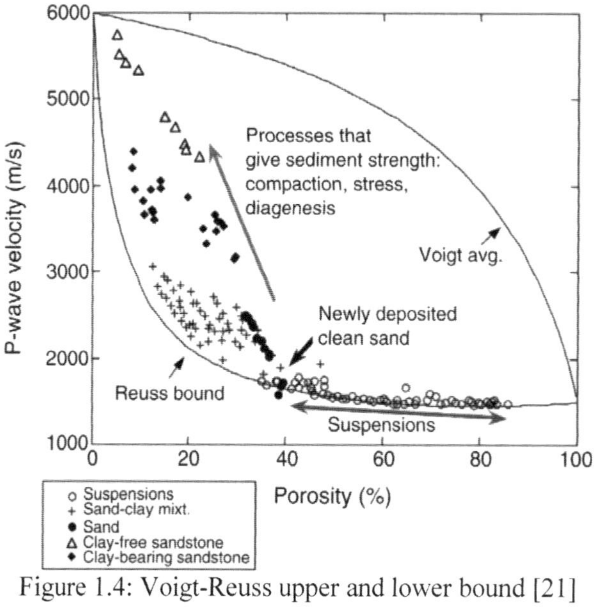

Figure 1.4: Voigt-Reuss upper and lower bound [21]

Fluid bulk modulus and fluid density will manifest the fluid substitution changes and therefore, they will affect the model and define the variations in parameters for seismic modelling. To establish matrix characteristics, one that needs to determine the matrix bulk modulus (Kbulk). These values are tabulated for main mineral components such as quartz and calcite among others. Most of the time, the matrix is composed of different kinds of minerals, and in case that fractional content information is available, the bulk modulus can be calculated using Voigt-Reuss-Hill (VRH) equation [22]. It is possible to obtain the bulk modulus (Kvrh) using Voigt and Reuss bulk modulus respectively (Kvoigt and Kreuss) [18]:

$$K_{bulk} = \frac{1}{2}(K_{voigt} + K_{reuss}) \quad (1.2)$$

$$K_{reuss} = \left[\frac{F_1}{K_1} + \frac{F_2}{K_2}\right]^{-1} \quad (1.3)$$

$$K_{voigt} = [F_1K_1 + F_2K_2] \quad (1.4)$$

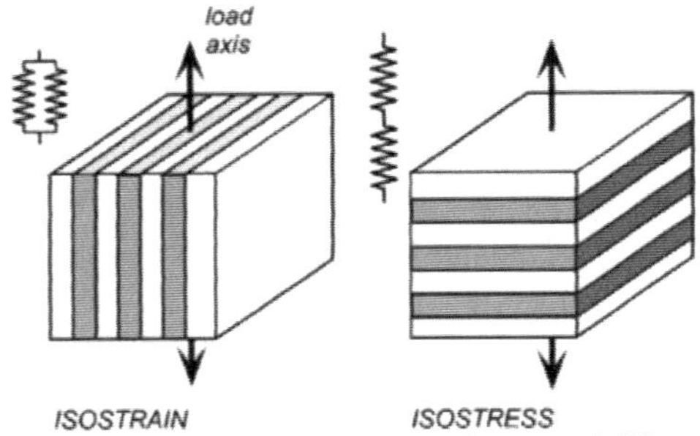

Figure 1.5: Voigt (isostrain) and Reuss (isostress) illustration [22].

2. Methodology

2.1 Homogeneous Core Simulation

To study the P-wave velocity, amplitude and resistivity variation when CO2 is injected to sandstone sample, we use computer simulation based on cylindrical sample of Berea sandstone (50 mm in diameter and 100 mm in length, 19% porosity) [10] as a reference value.

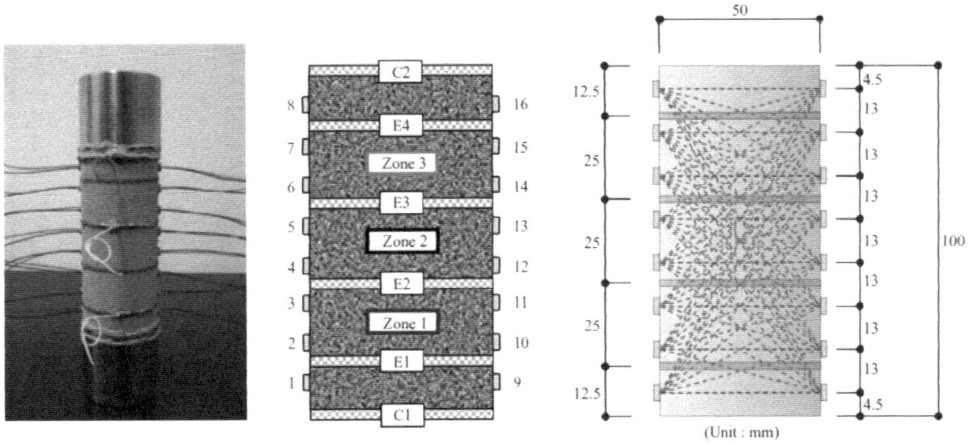

Figure 2.1: Configuration of piezoelectric transducers and electrodes with Berea sandstone sample for seismic velocity and resistivity measurements, respectively. The figure indicates the zone utilized for resistivity measurements as well as a diagram of ray paths for P-wave velocity measurements and P-wave velocity tomography [10].

Two flat current electrodes, 50mm diameter, were attached at the top and the bottom of the sample, and four ring-shaped potential electrodes of 2.5mm width were attached to the sample's surface at a spacing of 25 mm and made as an array. These electrodes are made of copper mesh coated with silver, which to avoid corrosion due to contact with acid.

For seismic value monitoring, our equipment setting were different from [10]. We use a coupled transducer which will be source and receiver on both end of the sample. To have a more realistic data value with reservoir condition, the experimental setup is made as in [10]. Note that the Berea sandstone will be in water saturated and later in oil saturated condition. The measured value (for water-CO2 fluid substitution) will be cross checked with values from Vera [23].

CO2 injection is monitored in time and saturation, and the resultant changes in resistivity and P-wave and S-wave values will be recorded. At the end of the experiments, we wish to obtain updated experimental result as in [10] by computer simulation for homogeneous media.

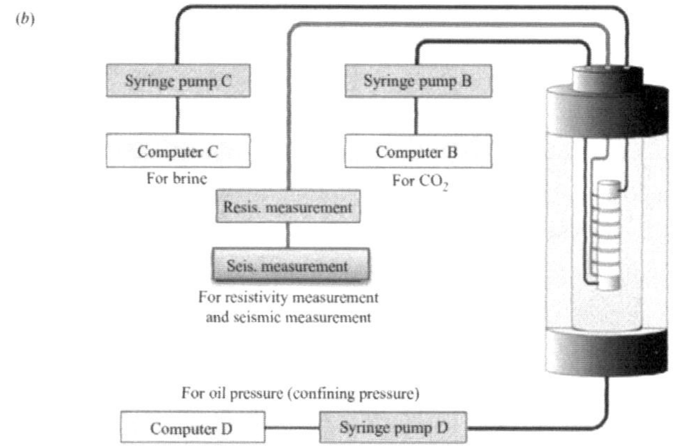

Figure 2.2: Configuration of experimental setup (a) and measurement setup (b) [10]

For the computer simulation, we developed a Python language programmed using hybrid Python version IDE, named Anaconda, first for homogeneous media. The purpose of the simulation is to generate expected result, with total control on the rock properties parameter, to enable us to measure variable range of rock properties rather than to rely on a single core sample.

The following setup is for 4 layers modelling:

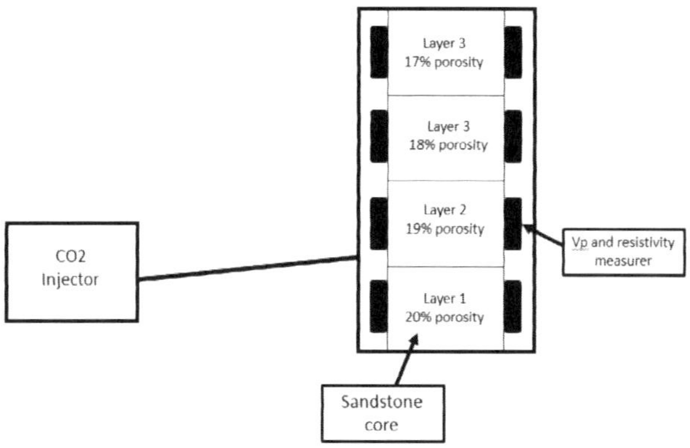

For the purpose of our experiment, based on [10] we developed a core model with parameters as follows:

Table 2.1: Core model parameters

Parameters	Sandstone	Brine	Gas
Bulk Modulus (GPA)	39	3.0	0.038
Shear Modulus (GPA)	44	-	-
Density (g/cc)	2.65	1.039	0.15
Resistivity (ohm.m)	15	1.0	30
Cementation Factor (m)	1.59	-	-
Biot Coefficient	2.0	-	-

The volume of 100% CO_2 saturation in the core was set to be 20 mL for each layer, with rate of injection of 0.005 mL/min. The first layer to second layer is 50 min after. The numerical methods to calculate the velocity and resistivity of the simulation are as follows:

Vp calculation:

1. Calculation of matrix modulus by Krief et al. method.
2. Calculation of dry and brine saturated formations by Biot-Gassmann formula.
3. Mix the gas and brine fluids modulus based on their saturation by using Wood's formula to get effective fluid modulus.
4. Repeat Biot-Gassmann and use effective fluids calculated.

Resistivity calculation

1. By matrix modulus, gain porosity extrapolation from brine velocity by Gassmann formula for each brine saturation percentage.
2. Calculate brine saturated resistivity from CRIM formula.
3. By brine modulus, repeat porosity extrapolation from mix fluids (brine-gas) velocity by Gassmann formula.
4. Using brine saturated resistivity, repeat CRIM formula to calculate mix fluids saturated resistivity.

3. Result and Discussion

3.1 Core simulation results

We acquire the simulation result using Anaconda programming IDE, based on Python 2.7 architecture. The results of the heterogeneous core sampling and comparison with our simulation (homogeneous media) are as follows.

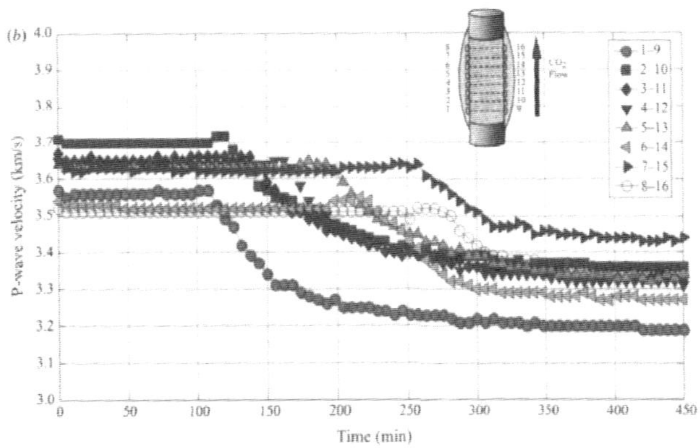

Figure 3.1: Time variation of velocity by injection of supercritical CO_2 into water saturated Berea Sandstone in heterogeneous media [10]

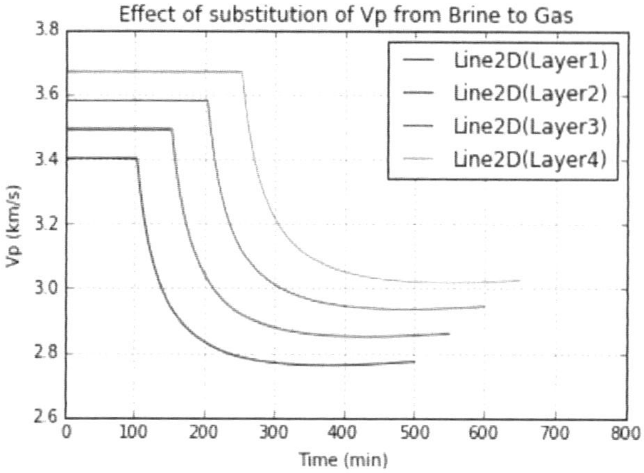

Figure 3.2: The effect of substitution of Vp from brine to CO2 gas in homogeneous media simulation.

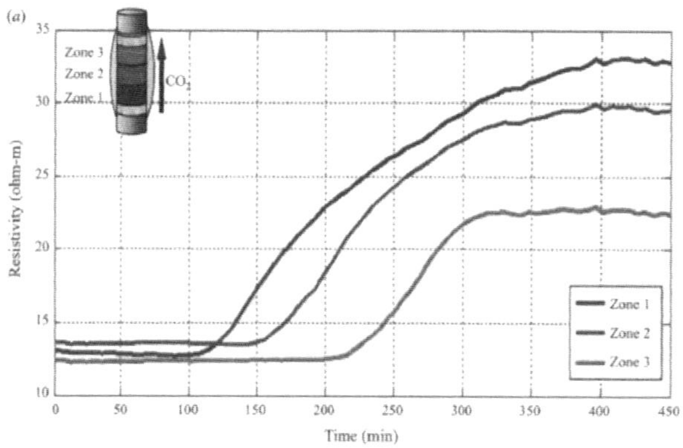

Figure 3.3: Time variation of resistivity by injections of supercritical CO2 into water saturated Berea sandstone in heterogeneous media [10]

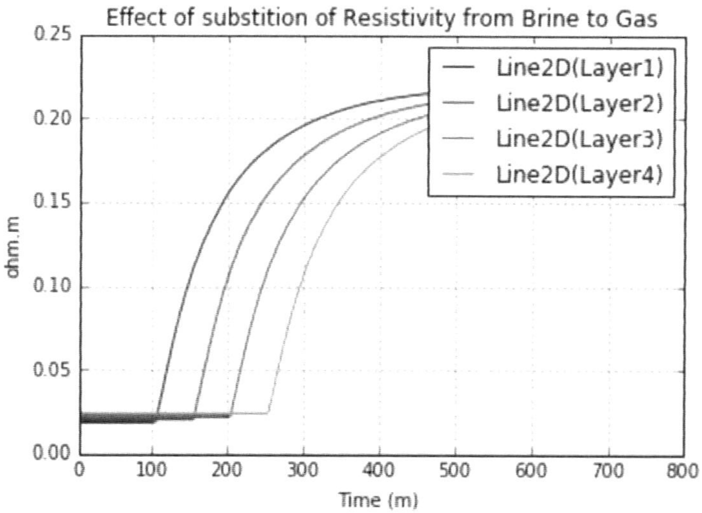

Figure 3.4: Effect of substitution of resistivity from brine to gas in homogeneous media simulation.

Our simulation shows the same trend with established work by [10] although our simulation work is applied on homogeneous media while Kim et al. [10] work is applied on core sample which has heterogeneity included. Therefore, we can conclude that our simulation method can be used in further work to monitor CO2 injection in sandstone using seismic and resistivity method. By having a computer simulation method, we can have a wide range of prediction on the CO2 effect on seismic as well as resistivity value. Rather than to rely on core sampling in laboratory, we can now predict the expected value of seismic and resistivity in a controlled environment with a low cost, fast and robust method.

4. Conclusions

Despite seismic data quality improvements and excellent well control, to achieve correct interpretation, rock physics is required to quantify the connection between geology and seismic data [15]. It is noted from [10] that the seismic sensitivity in detecting CO2 will be not significant after it reach 20% saturation, therefore resistivity value which remain in steady trend of increment will be an alternative once more CO2 volume is added to the reservoir. We have demonstrated that our simulation which is first done for homogeneous media will be a great tool in expecting the outcome in monitoring CO2 EOR using seismic and resistivity. However, we need to extend the research towards heterogeneous media simulation as the real geology often seen as heterogeneous.

References

1. Arts, R., et al., *Ten years' experience of monitoring CO2 injection in the Utsira Sand at Sleipner, offshore Norway.* First break, 2008. **26**(1).
2. Arts, R. and V. Vandeweijer, *The challenges of monitoring CO 2 storage.* The Leading Edge, 2011. **30**(9): p. 1026-1033.
3. Couëslan, M., et al., *Monitoring CO2 injection for carbon capture and storage using time-lapse 3D VSPs.* The Leading Edge, 2013. **32**(10): p. 1268-1276.
4. Juanes, R., et al., *Impact of relative permeability hysteresis on geological CO2 storage.* Water Resources Research, 2006. **42**(12).
5. Chapel, D.G., C.L. Mariz, and J. Ernest. *Recovery of CO2 from flue gases: commercial trends.* in *Canadian Society of Chemical Engineers Annual Meeting.* 1999.
6. Gozalpour, F., S. Ren, and B. Tohidi, *CO2 EOR and storage in oil reservoir.* Oil & gas science and technology, 2005. **60**(3): p. 537-546.
7. Berger, B., O. Kaarstad, and H.A. Haugen. *Creating a large-scale CO2 infrastructure for enhanced oil recovery.* in *Proceedings of the 7th International Conference on Greenhouse Gas Control Technologies, Vancouver, Canada. http://uregina.ca/ghgt7/PDF/papers/nonpeer/108. pdf.* 2004.
8. Archie, G.E., *The electrical resistivity log as an aid in determining some reservoir characteristics.* Transactions of the AIME, 1942. **146**(01): p. 54-62.
9. Park, J., et al. *Feasibility Study on Seismic and CSEM Monitoring of CO2 Injection Based on Laboratory Acoustic and Resistivity Measurement.* in *The Third Sustainable Earth Sciences Conference & Exhibition.* 2015.
10. Kim, J., T. Matsuoka, and Z. Xue, *Monitoring and detecting CO2 injected into water-saturated sandstone with joint seismic and resistivity measurements.* Exploration Geophysics, 2011. **42**(1): p. 58-68.
11. Frohlich, R.K. and C.D. Parke, *The electrical resistivity of the vadose zone—field survey.* Groundwater, 1989. **27**(4): p. 524-530.
12. JafarGandomi, A. and A. Curtis, *Detectability of petrophysical properties of subsurface CO 2-saturated aquifer reservoirs using surface geophysical methods.* The Leading Edge, 2011. **30**(10): p. 1112-1121.
13. Lumley, D.E., *Time-lapse seismic reservoir monitoring.* Geophysics, 2001. **66**(1): p. 50-53.

14. Saito, H., et al., *Time-lapse crosswell seismic tomography for monitoring injected CO2 in an onshore aquifer, Nagaoka, Japan.* Exploration Geophysics, 2006. **37**(1): p. 30-36.
15. Avseth, P., T. Mukerji, and G. Mavko, *Quantitative seismic interpretation: Applying rock physics tools to reduce interpretation risk*. 2005: Cambridge university press.
16. Sokolnikoff, I.S. and R.D. Specht, *Mathematical theory of elasticity*. Vol. 83. 1956: McGraw-Hill New York.
17. Han, D.-h. and M.L. Batzle, *Gassmann's equation and fluid-saturation effects on seismic velocities.* Geophysics, 2004. **69**(2): p. 398-405.
18. Smith, T.M., C.H. Sondergeld, and C.S. Rai, *Gassmann fluid substitutions: A tutorial.* Geophysics, 2003. **68**(2): p. 430-440.
19. Carcione, J.M., et al., *Cross-hole electromagnetic and seismic modeling for CO2 detection and monitoring in a saline aquifer.* Journal of Petroleum Science and Engineering, 2012. **100**: p. 162-172.
20. Batzle, M. and Z. Wang, *Seismic properties of pore fluids.* Geophysics, 1992. **57**(11): p. 1396-1408.
21. Watt, J.P., *Hashin-Shtrikman bounds on the effective elastic moduli of polycrystals with orthorhombic symmetry.* Journal of Applied Physics, 1979. **50**(10): p. 6290-6295.
22. Hill, R., *Elastic properties of reinforced solids: some theoretical principles.* Journal of the Mechanics and Physics of Solids, 1963. **11**(5): p. 357-372.
23. Vera, V.C., *Seismic modelling of CO2 in a sandstone aquifer, Priddis, Alberta*. 2012, University of Calgary.

Application of Simultaneous Inversion Method to Predict the Lithology and Fluid Distribution in "X" Field, Malay Basin

Ashley Aisyah Yoong[1], Luluan Almanna Lubis[1], Deva P. Ghosh[1]
Geoscience Department, Universiti Teknologi PETRONAS, 32610 Tronoh, Perak, Malaysia.

E-mail: ashleyaisyahyoong@gmail.com

Abstract: Reservoir characterization is one of the most crucial processes in the exploration of hydrocarbons. Thus, having sufficient information such as the lithology and fluid distribution of a reservoir is vital. With the application of Simultaneous Seismic Inversion, we will be able to improve the prediction of the lithology and fluid distribution of the prospect area, as it generates information such as Density (ρ), Lambda-rho (λρ) (Incompressibility), and Mu-rho (μρ) (Rigidity). Real data from the Malay basin were used to show the effectiveness of this method in enhancing the resolution and characterizing the gas sand layer of the prospect.

1. Introduction

Seismic reflection methods have been widely used over the years to locate potential oil and gas reservoirs, as well as providing information on the physical properties of the reservoir rocks. Changes in physical rock properties such as density, results in the change of seismic acoustic impedance, causing a significant effect that could be observed in a high quality seismic data.

Back in the years, reservoir characterization has only been using Acoustic Impedance (P-impedance) seismic data to obtain information. However, this approach is insufficient, since P-impedance seismic data is unable to distinguish the effects between lithology and fluid content. Fortunately, we could overcome this limitation of interpretation by extracting information from the Shear Impedance (S-impedance) seismic data that highlights the fluid content of an area.

This study seeks to demonstrate the application of the Simultaneous Seismic Inversion method on the "X" field which will enhance the prediction of the lithology and fluid distribution in the prospect area. This method estimates the P and S-impedance by inverting each partial angle stack data simultaneously using the extracted wavelet from each angle stack. The inversion results such as Density (ρ), Lambda-rho (λρ) *(Incompressibility)*, and Mu-rho (μρ) *(Rigidity)* are proven to be useful for reservoir characterization of a prospect area.

[1] Geosciences Department, Universiti Teknologi PETRONAS, Malaysia, Bandar Seri Iskandar, 32610, Tronoh, Perak Darul Ridzuan, Malaysia
[2] Petroleum Engineering Department, Universiti Teknologi PETRONAS, Malaysia, Bandar Seri Iskandar, 32610, Tronoh, Perak Darul Ridzuan, Malaysia

IOP Conf. Series: Earth and Environmental Science **38** (2016) 012007 doi:10.1088/1755-1315/38/1/012007

2. Methodology

2.1 3D Seismic Data

For this project, the available seismic data are only post-stack angle volumes also known as angle stack. The seismic survey coverage ranges from Inline 980 to 1582 and X-line 2100 to 6090 for each volume. Three angle stacks (post-stack) are used, where each stack represents a different range of angles as shown in Figure 4.13.

a. Near stack (5° - 15°)
b. Mid stack (15° - 30°)
c. Far stack (30° - 45°)

These stacks are loaded separately as post-stack data into the Hampson Russell (HRS) software, before creating pseudo gathers (pre-stack) by extracting a trace from each volume for every offset. In other words, each offset has three different angle traces (15°, 30° and 45°). This seismic loading workflow definitely differs from the conventional workflow of simultaneous inversion, where we create pre-stacks volume from given post-stack volumes.

2.2 Well data

A complete well log data per well is needed before we could proceed with the simultaneous inversion. The essential data required for the inversion workflow are P-wave sonic, S-wave sonic , density and checkshots logs.

2.3 Method

The cross plot approach is being applied on the X-3 well, in order to determine the feasilibility of the simultaneous seismic inversion process on the dataset, before proceeding to the full inversion workflow. Figure 4.2 represents the cross plot of Vp/Vs ratio versus P-impedance at the gas sand zone of interest with a colour scale that indicates the gamma ray values. From the cross plot constructed, it can be observed that there is a separation between the points scattered, making it possible to differentiate the gas sand from the brine sand. Therefore, the inversion workflow can be proceed.

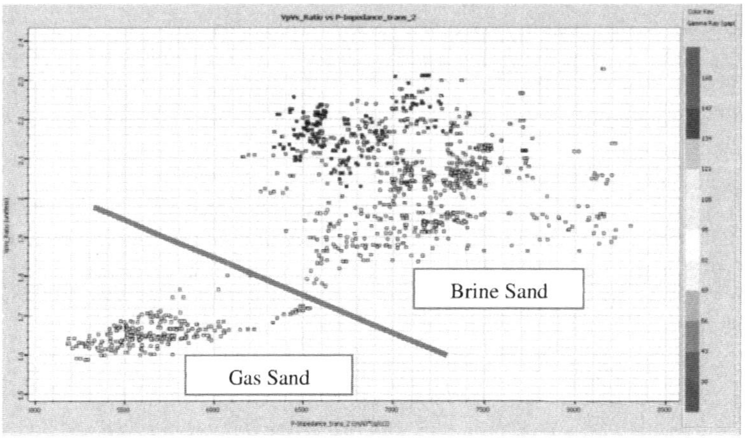

Figure 2.1 Cross plot of Vp/Vs ratio against P-impedance

For a pre-stack data, it is highly expected that the high frequency energy of the seismic data would experience a continuous loss as it moves from the near to the far offset. Thus, each trace of the CDP gather would have a different wavelet representation. However, it is believed that extracting a wavelet for every 15 degrees of incident angle is adequate to be used for the inversion process, instead of extracting for each offset, which is not practical yet time consuming.

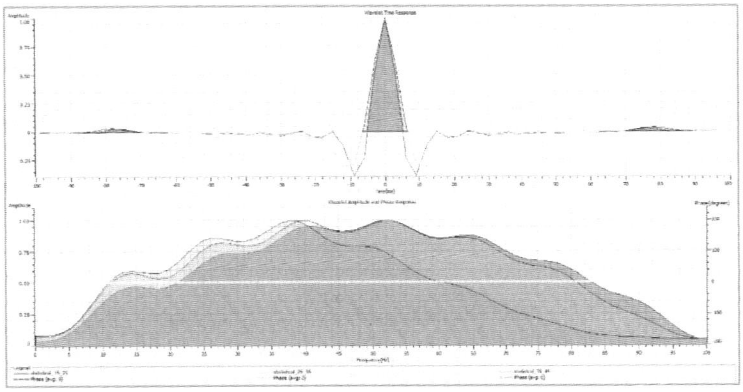

Figure 2.2 Extraction of Statistical wavelet

Synthetic traces are the ultimate key in the well-tie process, where well data are linked to the seismic data. In the process of creating synthetic traces, a bandpass filter is applied to the well log curves, in order to obtain the same frequency range as the seismic data. A filter with a low pass of 0Hz and a high pass of 50 Hz applied to the log curves, permits us to observe how well the inversion has worked within the seismic band, since the end result of inversion are expected to be in that particular range of frequency.

When both wavelet and the low frequency model have been obtained, the convolution process is carried out and applied to the volume. As a result, three volumes is generated namely, inverted P-impedance (Zp), inverted S-impedance (Zs) and density.

The inversion workflow continues by transforming the output volumes into Lambda-rho, Mu-rho and Vp/Vs volumes. The Lambda-mu-rho approach was proposed by Goodway et al. in 1997, where he introduced the Lame parameters (λ and μ) and density ρ as the core parameters of the Simultaneous Inversion new approach.

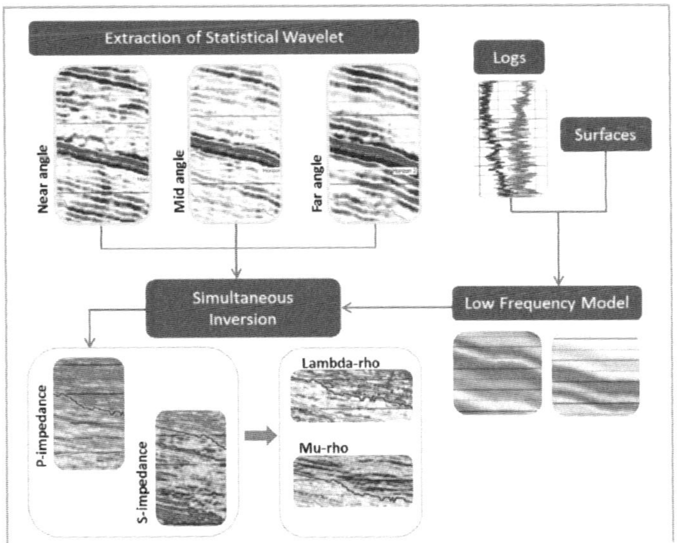

Figure 2.3 The Simultaneous Inversion Workflow. (Edited from Jason, CGG).

3. Result and Discussion

Pre Inversion Analysis acts as a platform for geophysicists to test a variety of inversion parameters on a selected well before applying to the seismic volume. It permits us to measure the success of the inversion, as well as to analyse different possible outcomes resulting from different parameters

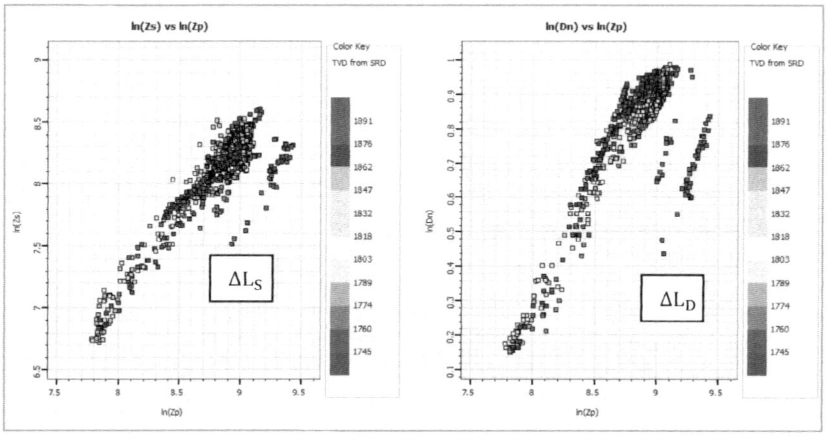

Figure 3.1 Cross plot Inversion analysis at wells.

Both cross plot of $\ln Z_S$ vs $\ln Z_P$, and $\ln \rho$ vs $\ln Z_P$, could be consider as having a directly proportional trend which suggests that the inversion workflow is accurate and no shifting will be needed.

Apart from the direction of the trend, the plotted points do give other insights too. In Figure 3.1, we could observe that each case has two trend lines being plot. The line with the most plotted data would represent the best fit line, while the deviation of data away from this straight line, are believed to be fluid anomalies, ΔL_S and ΔL_D.

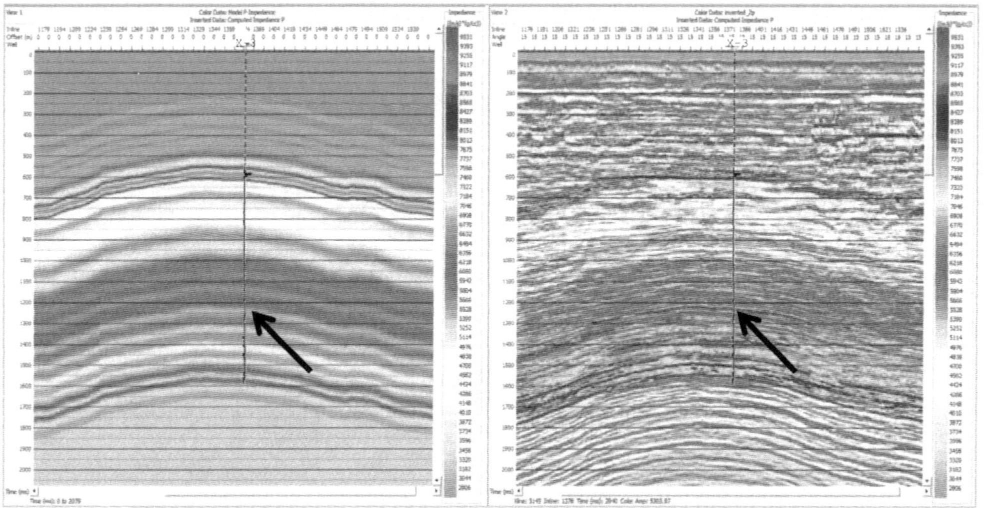

Low Frequency P-impedance model Inverted Zp model

Figure 3.2 Comparison between the P-impedance and the Inverted Zp model at X-3.

Once the inversion is applied to the volume, we could evaluate that the inversion results have a better image with clearer and more accurate impedance model. The arrows in Figure 4.24 showing the clear separation of top gas sand (higher impedance - purple) in the inverted Zp model compared to the P-impedance model.

Lambda-rho and mu-rho are indeed powerful attributes in characterizing a reservoir. Lambda-rho which is sensitive to the pore fluids, shows a smaller thickness in Figure 3.3 compared to the Mu-rho in Figure 3.4 . Since mu-rho is only sensitive to the rock matrix, thus its responses characterize the rigidity of the gas sand layer, rather than the pore fluid content. Meanwhile the Lambda-rho gives the distribution in pore fluid within the gas sand layer. By integrating the information, we would be able to predict the actual condition of the gas sand layer, directly reduces the risk of drilling a dry well .

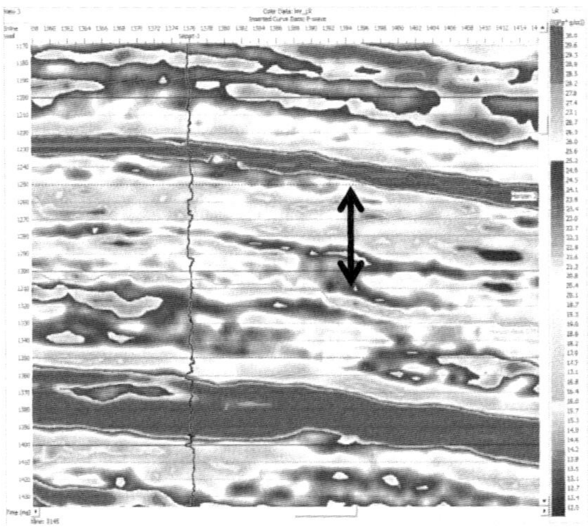

Figure 3.3 Lambda-Rho Volume at X-3.

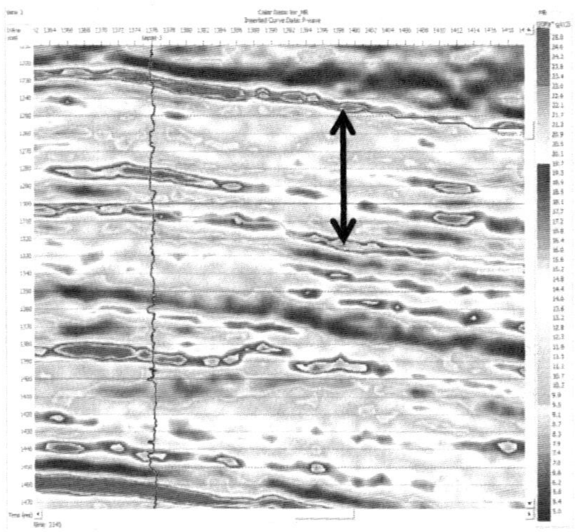

Figure 3.4 Mu-Rho Volume at X-3.

4. Conclusions

We have demonstrated that the Simultaneous Seismic Inversion is indeed a powerful tool in predicting the lithology and fluid distribution in the "X" field gas sand layer. The resolution achieved through this inversion allows us to characterize a reservoir better, especially by applying the Lambda-mu-rho analysis onto the inversion volumes.

5. References

Buland, A., & Omre, H. (2003). Bayesian linearized AVO inversion. *Geophysics*, 185-198.

Castagna, J., Batzle, M., & Eastwood, R. (1985). Relationship between compressional-wave and shear wave velocities in clastic silicate rocks. *Geophysics*, 571-581.

Fatti, J., Smith, G., Vail, P., Strauss, P., & Levitt, P. (1994). Detection of gas in sandstone reservoirs using AVO analysis: a 3D Seismic Case History Using the Geostack Techniue. *Geophysics*, 1362-1376.

Francis, A. (2014). *A Simple Guide to Seismic Inversion*. Retrieved from GEOEXPRO: http://www.geoexpro.com/articles/2014/06/a-simple-guide-to-seismic-inversion

Gardner, G. H., Gardner, L., & Gregory, A. (1985). Formation velocity and density - The diagnostic basics for stratigraphic traps. *Geophysics*, 2085-2095.

Goodway, B., Chen, T. and Downton, J. (1997). Improved AVO fluid detection and lithology discrimination using Lamé petrophysical parameters, 67th *Ann. Internat. Mtg: SEG*, 183-186.

Hampson, D. P., Russell, B., & Bankhead, B. (2006). Simultaneous Inversion of Pre-stack Seismic Data. *Geohorizons*, 13-17.

Jain, P. K., Jambhekar, C. V., & Pandey, P. B. (n.d.). Indentification of gas using Vp/Vs vis-a-vis Poisson's ratio. *9th Biennial International Conference & Exposition on Petroleum Geophysics.*

Lubis, L.A. (2014). Seismic Inversion [PowerPoint Slides]. Rock Physics and AVO Analysis.

Pendrel, J., Debeye, H., Pedersen-Tatalovic, R., Goodway, B., Dufour, J., Bogaards, M., et al. (1999). Estimation and interpretation of P and S impedance volumes from simultaneous inversion of. *Research Gate.*

Simmons, J., & Backus, M. (1996). Waveform-based AVO inversion and AVO prediction-error. *Geophysics*, 1575-1588.

IOP Conf. Series: Earth and Environmental Science **38** (2016) 012008 doi:10.1088/1755-1315/38/1/012008

The ranks of Indonesian and Japanese industrial sectors

Ubaidillah Zuhdi

Faculty of Management and Economics, Gdansk University of Technology, Gdansk 80-233, Poland

zuhdi@zie.pg.gda.pl

Abstract. The purpose of this study is to determine the ranks of Indonesian and Japanese industrial sectors from the economic point of view. The analysis period of this study is 2005. This study employs one of the well-known analysis tools in the economic topic, the Input-Output (IO) analysis. More specifically, this study uses the analysis methods in the IO analysis, backward and forward linkages, in order to achieve the purpose. The results of calculations show that the orders of the ranks depend on the method used. Nevertheless, from the results, one can say that the manufacturing industry was a leading sector in the Indonesian economy on the analysis period. On the other hand, for the Japanese case, the sector which had the beneficial effects in the Japanese economy on the analysis period was the transport.

1. Introduction

An analysis of the industrial sectors of one country will continually give the insights about their characteristics. These insights will be the triggers in order to improve the sectors. Further, the improvement actions for the industries might increase the macroeconomics conditions of the country.

The examples of previous studies which focuses on the analysis are [1], [2], [3], [4], [5], and [6]. To the best of my knowledge, after learning the studies, the research which the subject is to determine the ranks of the industrial sectors of countries is still needed. The research is required in order to get the understanding about the beneficial sectors in the economic activities of focused countries. This study tries to fulfill the need.

The purpose of this study is to determine the ranks of Indonesian and Japanese industrial sectors from the economic point of view. The analysis period of this study is 2005. This study employs one of the well-known analysis tools in the economic discussion, the Input-Output (IO) analysis, in order to achieve the purpose. One thing that all readers should notice is that this study focuses on the observation rather than the comparison for the conditions of Indonesian and Japanese industrial sectors.

2. Methodology

The methodology of this study is described as follows. The first step is to explain the data sources. The data sources of this study are 2005 Indonesian and Japanese IO tables. The table of Indonesia is obtained from [7] while the Japanese one is acquired from [8]. The second step is to describe the industrial sectors of Indonesia and Japan used in this study. Tables 1 and 2 elaborate these sectors for the cases of Indonesia and Japan, respectively.

Table 1. Indonesian industrial sectors used in this study.

Sector Number	Sector Name
1	Agriculture, livestock, forestry, and fishery
2	Mining and quarrying
3	Manufacturing industry
4	Electricity, gas, and water supply
5	Construction
6	Trade, hotel, and restaurant
7	Transport and communication
8	Financial, real estate, and business services
9	Services

Table 2. Japanese industrial sectors used in this study.

Sector Number	Sector Name
1	Agriculture, forestry, and fishery
2	Mining
3	Manufacturing
4	Construction
5	Electricity, gas, and water supply
6	Commerce
7	Finance and insurance
8	Real estate
9	Transport
10	Information and communications
11	Public administration
12	Services
13	Activities not elsewhere classified

The third step is to conduct the calculations in order to determine the ranks of the industries of analyzed countries on the analysis period. The methods of backward and forward linkages, the analysis devices in the IO analysis, are employed in the calculations. Both linkages describe the relationship between a specific industrial sector and other industries. More specifically, the backward linkage focuses on the demands of the particular industry from other industrial sectors. These demands appear as a consequence of the industry as a purchaser in the economy. On the other hand, the forward linkage explains the consequence of the industrial sector as a seller.

The methods are suitable for determining the ranks of industrial sectors on the specific time period. [9] reaffirm this argument through the following statement:

"*Measures have been proposed to quantify such backward and forward linkages, or economic "connectedness." Comparisons of the strengths of backward and forward linkages for the sectors in a single economy provide one mechanism for identifying "key" or "leading" sectors in that economy (those sectors that are most connected and therefore, in some sense, most "important") and for grouping sectors into spatial clusters.*"

One can say that if the backward linkage value of sector *a* is higher than that of sector *b*, then sector *a* gives more beneficial effect because of the higher capability of the sector in attracting the economy through its demands. The similar argument which uses the supplier's point of view can be applied in analyzing the forward linkage values of two or more industries.

This study uses two forms for each linkage, namely "direct" and "total". These forms can be seen on both demand-side and supply-side IO models. [10] propose the concepts of the former form on both models. To summarize, this form only indicates the straightforward impacts of the demands and supplies of one specific industry. Further, [9] explain the concept for the backward linkage on the form using the following equation:

$$BL(d)_j = \sum_{i=1}^{n} a_{ij} \quad (1)$$

where $BL(d)_j$, a_{ij}, and n are the direct backward linkage of sector j, the direct input coefficient, and the number of industrial sectors, respectively. On the other hand, [11] describes the concept for the forward linkage on the form through the following equation:

$$F(d)_i = \sum_{j=1}^{n} a_{ij} \tag{2}$$

where $F(d)_i$ is the direct forward linkage of sector i. Both equations use the demand-driven IO model as the reference. The using of row sums in explaining the direct forward linkage of the specific sector in the model is also revealed by [9].

The insight of the "total" form for the backward linkage in the demand-driven IO model is proposed by Rasmussen (1957) in [9] while [9] describes the one for the forward linkage. To summarize, this form explains the straightforward and indirect effects of the demands and supplies of one particular industry. [9] explore the former insight by using the following equation:

$$BL(t)_j = \sum_{i=1}^{n} l_{ij} \tag{3}$$

where $BL(t)_j$ and l_{ij} are the total backward linkage of sector j, and Leontief inverse or the total requirements matrix, respectively. The latter insight, on the other hand, is explained by [11] by using the following equation:

$$F(d+i)_i = \sum_{j=1}^{n} \alpha_{ij} \tag{4}$$

where $F(d+i)_i$ and α_{ij} are the total forward linkage of sector i and Leontief inverse matrix, respectively. As with equations (1) and (2), the reference of equations (3) and (4) is also the demand-driven IO model.

[9] mention that the Ghosh inverse is suggested as a better tool in describing the total forward linkage. As a consequence, the supply-driven IO model is a suitable device for measuring the forward linkage. [9] describe the following equations in representing the linkage using the model:

$$FL(d)_i = \sum_{j=1}^{n} b_{ij} \tag{5}$$

$$FL(t)_i = \sum_{j=1}^{n} g_{ij} \tag{6}$$

where $FL(d)_i$, $FL(t)_i$, b_{ij}, and g_{ij} are the direct forward linkage of sector i, the total forward linkage of sector i, the coefficients matrix in a supply-driven IO model, and the Ghosh inverse, respectively. In this study, all above equations are used in the calculations step. The next step is to analyze the ranks of Indonesian and Japanese industrial sectors on the analysis period. Conclusions of this study, and suggestions for further researches are described in the final step.

3. Results and analysis

Tables 3 and 4 explain the ranks of Indonesian industrial sectors on the analysis period which are viewed from the results of calculations results using direct and total backwards linkages, respectively. On the other hand, tables 5 and 6 describe the ranks which are viewed from the results using direct and total forwards linkages, respectively. Meanwhile, the results using the supply-driven IO model for the Indonesian case are explored in tables 7 and 8. The analysis of the case then focuses on the top five sectors of each result. More specifically, one can argue that the sector number 3, manufacturing industry, was a leading sector in the Indonesian economy in 2005 because it appears as one of the top five sectors in tables 3 – 8. In other words, on the analysis period, the sector had beneficial effects in the Indonesian economy.

Tables 9 and 10 describe the ranks of Japanese industrial sectors on the analysis period which are viewed from the results of calculations using direct and total backwards linkages, respectively. Meanwhile, tables 11 and 12 explore the ranks which are viewed from the results using direct and total forwards linkages, respectively. On the other hand, the results using the supply-driven IO model for the Japanese case are explained in tables 13 and 14. As with the previous case, the top five sectors on each result are viewed in the analysis of the Japanese issue. Based on the analysis, one can say that the sector number 9, transport, was a leading sector in the Japanese economy in 2005 because it appears as one of the top five sectors in almost all tables which explain the ranks. In other words, on the analysis period, the sector had beneficial effects in the Japanese economy.

Table 3. The ranks of Indonesian industrial sectors on the analysis period, based on the results of calculations using equation (1).

Sector Number	Sector Name	Calculated Value
4	Electricity, gas, and water supply	0.697
5	Construction	0.642
3	Manufacturing industry	0.626
7	Transport and communication	0.512
9	Services	0.460
6	Trade, hotel, and restaurant	0.407
8	Financial, real estate, and business services	0.320
1	Agriculture, livestock, forestry, and fishery	0.235
2	Mining and quarrying	0.181

Table 4. The ranks of Indonesian industrial sectors on the analysis period, based on the results of calculations using equation (2).

Sector Number	Sector Name	Calculated Value
4	Electricity, gas, and water supply	2.379
5	Construction	2.258
3	Manufacturing industry	2.161
7	Transport and communication	2.026
9	Services	1.919
6	Trade, hotel, and restaurant	1.766
8	Financial, real estate, and business services	1.585
1	Agriculture, livestock, forestry, and fishery	1.440
2	Mining and quarrying	1.295

IOP Conf. Series: Earth and Environmental Science **38** (2016) 012008 doi:10.1088/1755-1315/38/1/012008

Table 5. The ranks of Indonesian industrial sectors on the analysis period, based on the results of calculations using equation (3).

Sector Number	Sector Name	Calculated Value
3	Manufacturing industry	1.730
8	Financial, real estate, and business services	0.519
2	Mining and quarrying	0.380
6	Trade, hotel, and restaurant	0.354
7	Transport and communication	0.305
1	Agriculture, livestock, forestry, and fishery	0.256
9	Services	0.221
4	Electricity, gas, and water supply	0.207
5	Construction	0.110

Table 6. The ranks of Indonesian industrial sectors on the analysis period, based on the results of calculations using equation (4).

Sector Number	Sector Name	Calculated Value
3	Manufacturing industry	4.317
8	Financial, real estate, and business services	1.912
2	Mining and quarrying	1.798
1	Agriculture, livestock, forestry, and fishery	1.703
6	Trade, hotel, and restaurant	1.656
7	Transport and communication	1.559
9	Services	1.381
4	Electricity, gas, and water supply	1.320
5	Construction	1.183

Table 7. The ranks of Indonesian industrial sectors on the analysis period, based on the results of calculations using equation (5).

Sector Number	Sector Name	Calculated Value
8	Financial, real estate, and business services	0.803
4	Electricity, gas, and water supply	0.690
2	Mining and quarrying	0.685
1	Agriculture, livestock, forestry, and fishery	0.644
3	Manufacturing industry	0.594
7	Transport and communication	0.496
6	Trade, hotel, and restaurant	0.352
9	Services	0.218
5	Construction	0.086

Table 8. The ranks of Indonesian industrial sectors on the analysis period, based on the results of calculations using equation (6).

Sector Number	Sector Name	Calculated Value
8	Financial, real estate, and business services	2.518
2	Mining and quarrying	2.397
4	Electricity, gas, and water supply	2.393
1	Agriculture, livestock, forestry, and fishery	2.302
3	Manufacturing industry	2.095
7	Transport and communication	1.925
6	Trade, hotel, and restaurant	1.631
9	Services	1.414
5	Construction	1.167

Table 9. The ranks of Japanese industrial sectors on the analysis period, based on the results of calculations using equation (1).

Sector Number	Sector Name	Calculated Value
13	Activities not elsewhere classified	1.161
3	Manufacturing	0.698
2	Mining	0.570
4	Construction	0.538
9	Transport	0.522
5	Electricity, gas, and water supply	0.510
1	Agriculture, forestry, and fishery	0.472
10	Information and communications	0.409
12	Services	0.380
7	Finance and insurance	0.363
6	Commerce	0.315
11	Public administration	0.263
8	Real estate	0.146

Table 10. The ranks of Japanese industrial sectors on the analysis period, based on the results of calculations using equation (2).

Sector Number	Sector Name	Calculated Value
13	Activities not elsewhere classified	3.044
3	Manufacturing	2.649
4	Construction	2.224
2	Mining	2.168
9	Transport	2.081
1	Agriculture, forestry, and fishery	2.071
5	Electricity, gas, and water supply	2.063
12	Services	1.799
10	Information and communications	1.797
7	Finance and insurance	1.669
6	Commerce	1.600
11	Public administration	1.558
8	Real estate	1.279

Table 11. The ranks of Japanese industrial sectors on the analysis period, based on the results of calculations using equation (3).

Sector Number	Sector Name	Calculated Value
3	Manufacturing	1.614
7	Finance and insurance	1.028
12	Services	1.009
9	Transport	0.742
10	Information and communications	0.373
6	Commerce	0.358
11	Public administration	0.280
5	Electricity, gas, and water supply	0.269
2	Mining	0.175
1	Agriculture, forestry, and fishery	0.159
4	Construction	0.157
8	Real estate	0.113
13	Activities not elsewhere classified	0.071

Table 12. The ranks of Japanese industrial sectors on the analysis period, based on the results of calculations using equation (4).

Sector Number	Sector Name	Calculated Value
3	Manufacturing	5.249
12	Services	3.107
7	Finance and insurance	2.549
9	Transport	2.322
6	Commerce	1.828
10	Information and communications	1.751
5	Electricity, gas, and water supply	1.523
2	Mining	1.417
1	Agriculture, forestry, and fishery	1.321
11	Public administration	1.317
4	Construction	1.260
8	Real estate	1.227
13	Activities not elsewhere classified	1.133

Table 13. The ranks of Japanese industrial sectors on the analysis period, based on the results of calculations using equation (5).

Sector Number	Sector Name	Calculated Value
2	Mining	16.321
13	Activities not elsewhere classified	1.167
1	Agriculture, forestry, and fishery	0.825
7	Finance and insurance	0.709
5	Electricity, gas, and water supply	0.677
3	Manufacturing	0.647
9	Transport	0.641
10	Information and communications	0.581
12	Services	0.353
6	Commerce	0.346
4	Construction	0.144
8	Real estate	0.125
11	Public administration	0.029

Table 14. The ranks of Japanese industrial sectors on the analysis period, based on the results of calculations using equation (6).

Sector Number	Sector Name	Calculated Value
2	Mining	40.454
13	Activities not elsewhere classified	3.405
1	Agriculture, forestry, and fishery	3.014
7	Finance and insurance	2.556
9	Transport	2.544
3	Manufacturing	2.449
5	Electricity, gas, and water supply	2.437
10	Information and communications	2.139
12	Services	1.746
6	Commerce	1.735
4	Construction	1.263
8	Real estate	1.247
11	Public administration	1.098

4. Conclusions and further researches

This study determines the ranks of Indonesian and Japanese industries by using the analysis methods in the IO analysis, backward and forward linkages. The analysis period in this study is 2005. The results of calculations show that the orders of the ranks depend on the method used. Nevertheless, from the results, one can say that the manufacturing industry was a leading sector in the Indonesian economy on the analysis period. On the other hand, for the Japanese case, the sector which had the beneficial effects in the Japanese economy on the analysis period was the transport.

The ranks on the specific period can be seen from this study. Nevertheless, the deeper analysis regarding the results of calculations are needed in order to get the comprehensive depiction of the ranks on the period. This study suggests the analysis as a further research.

Another suggested further research from this study is to conduct the international comparison for Indonesia and Japan in the discussed topic. As described in Introduction, this study focuses on the observation rather than the comparison for the industrial sectors conditions of both countries on the specific period. The comparison will be an interesting further research because it can describe the comprehensive economic situations of both countries on the period. The other suggested further research is to expand the analysis period so the movement patterns of the industrial sectors of both countries can be explained.

References

[1] Zuhdi U 2015 An analysis of the role of information and communication technology sectors on Japanese national economy from 1995 through 2005: an application of multiplier analysis *IOP Conference Series: Earth and Environmental Science* **23** 012014

[2] Zuhdi U 2015 An application of multiplier analysis in analyzing the role of information and communication technology sectors on Indonesian national economy: 1990-2005 *IOP Conference Series: Earth and Environmental Science* **23** 012015

[3] Zuhdi U 2015 The dynamics of Indonesian creative industry sectors: an analysis using input-output approach *Journal of the Knowledge Economy* **6** 1177–90

[4] Zuhdi U 2015 An application of input-output analysis in analyzing the impacts of final demands changes on the total outputs of Japanese energy sectors: a further study *Journal of Physics: Conference Series* **622** 012041

[5] Zuhdi U 2014 The other perspective related to the role of information and communication technologies sectors in national economy: the case of Japan *Advanced Science Letters* **20** 483–6

[6] Zuhdi U 2014 Encouraging information and communication technology sectors using input-output approach: the case of Indonesia *Advanced Science Letters* **20** 199–202

[7] Statistics Indonesia 2015 *Statistical Yearbook of Indonesia 2015* [online] https://www.bps.go.id/website/pdf_publikasi/Statistik-Indonesia-2015.pdf (accessed 17 March 2016)

[8] Japanese Ministry of Internal Affairs and Communications 2009 *Introduction: The Japanese Economy and the 2005 Input-Output Tables* [online] http://www.soumu.go.jp/main_content/000327476.pdf (accessed 26 April 2016)

[9] Miller R E and Blair P D 2009 *Input-Output Analysis: Foundations and Extensions* (Cambridge: University Press)

[10] Chenery H B and Watanabe T 1958 International comparisons of the structure of productions *Econometrica* **4** 487–521

[11] Nazara S 2005 *Input-Output Analysis* [in Indonesian] (Jakarta: The Faculty of Economics of University of Indonesia)

IOP Conf. Series: Earth and Environmental Science **38** (2016) 012009 doi:10.1088/1755-1315/38/1/012009

Exploring the Opportunities of a Balloon-Satellite in Bangladesh for Weather Data Collection and Vegetative Analysis

Md. Ishraque Bin Shafique, M.A. Razzaq Halim, Fazle Rabbi and Md. Khalilur Rhaman

Abstract. For a third world country like Bangladesh, satellite and space research is not feasible due to lack of funding. Therefore, in order to imitate the principles of such a satellite Balloon Satellite can easily and inexpensively be setup. Balloon satellites are miniature satellites, which are cheap and easy to construct. This paper discusses a BalloonSat developed using a Raspberry Pi, IMU module, UV sensor, GPS module, Camera and XBee Module. An interactive GUI was designed to display all the data collected after processing. To understand nitrogen concentration of a plant, a leaf color chart is used. This paper attempts to digitalize this process, which is applied on photos taken by the BallonSat.

1. Introduction

The Space Age began with the development of space researches that begin in the mid nineteen hundred with the launch of Sputnik 1[1]. During that time, there was a competition going on mainly between USA and Russia. Soon, other countries had joined in this space race. Nowadays, space research is a very rich and emerging topic of science and engineering. Satellites play an important role in field of space study. These are man-made devices purposefully placed into the Earth orbit for numerous reasons. The cost of setting up a satellite is in the range of millions of dollars. For a developing country like Bangladesh, it is very difficult even to build a small satellite and without a practical satellite, space research will not be able to attract students [2]. *Table 1* shows the cost and the development time required to build several small-scale satellites [3].

Table 1:Small satellite projects and their development costs, courtesy of Delft University of Technology

Satellite	Country	Year	Mass (kg)	Development Time (years)	Development Cost (US Dollars)
UoSAT-1	Great Britain	1981	60	2.5	850,00.00
UoSAT-2	Great Britain	1984	65	0.5	1,300,000.00
Falcon Gold	United States	1997	179	0.7	550,00.00
SUNSAT	South Africa	1998	624	4.5	1,950,000.00
TUBSAT A	Germany	1991	35	5	2,000,000.00
BREMSAT	Germany	1994	63	3	3,500,000.00

Another cheap alternative to satellites is to build a Balloon Satellite. A BalloonSat is constructed using cheap, easily available off-the-shelf electronic components. The main reason behind making a BalloonSat is that students of developing countries can get initial idea and experience about space research with a very low budget [4]. The practical implementation of a BalloonSat, among others, might be to collect weather data periodically to figure out the changes in weather patterns. It can also be used to detect the rate of riverbank erosion [5]. Other uses might be by biologists to keep track of endangered animal species in a forest.

2. System Overview

2.1. Electronic Components

The main computational brain behind this BalloonSat is the Raspberry Pi 2 Model B+. It has been chosen for its high processing power with a clock speed of 900MHz and 1GB RAM. The Adafruit Ultimate GPS Module is used to keep track of the BalloonSat explores the skies with error less than 3m. A 10-degree Inertial Measurement Unit (IMU) is used in the setup. It senses 3 axes of acceleration, 3 axes of gyroscopic data, 3 axes of magnetic fields, pressure, temperature and altitude. In addition to these, a SI1145 sensor chip enables UV index sensing along with, IR and the visible light spectrum. The 5-megapixel Pi-Cam is programmed manually to take high definition photographs at 30s intervals, the circuit setup is as shown in *figure 1* and the actual circuit on PCB in *figure 2* both of which is our own design. The data collected by these sensors (except images) are then transmitted wirelessly using an XBee radio module. A 9V DC battery is regulated to 5V to power up the whole circuitry.

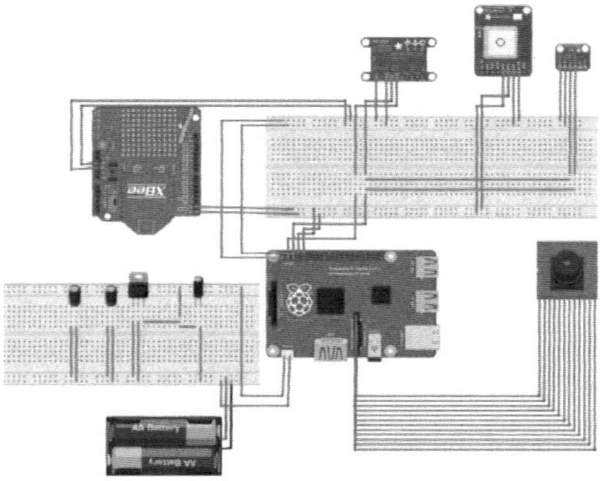

Figure 1: Circuit diagram of BalloonSat.

Figure 2: Actual circuit integration

2.2. The Ground Station (GS)

A Ground Station (GS) is developed to receive and store sensory data. The Graphical User Interface (GUI) of the GS has been coded using LabVIEW, which is a system-design platform and development environment for a visual programming language from National Instruments. As LabVIEW uses visual language, it is very easy to implement and build the GUI. The interface is grouped into several tabs where data from the sensors are viewed separately. The GS also has a data logging option enabled. Every time the program is run, it will save the data in an Excel spreadsheet for further analysis.

In the GUI, the tabs are labelled as Connection, GPS, Gyro, IMU Data, Light Sensor and Weather. The connection tab deals with the serial connection and shows the input in the com port. There is also a drop down list to select the correct com port to begin the connection. It also has a slider to adjust the delay of the receiving data. GPS tab shows the satellite time and date and also the longitude and latitude of the current position. There is also a button which will call up an executable file (.exe) which can plot the positions in a map. The Gyro tab shows the X, Y and Z components of the gyroscope in a live graph. The IMU Data (inertial measurement unit data) tab displays the pitch, roll, heading and altitudes in radial dials and a bar. The following tab is the Light Sensor tab which displays information from the SI1145 UV sensor. This tab has three dials to visualize Visible Light, Infrared and UV Index.

Finally, the Weather tab shows atmospheric temperature and pressure as well as the CPU temperature of the Raspberry Pi. *Figure 3* shows some of the tabs of the GUI.

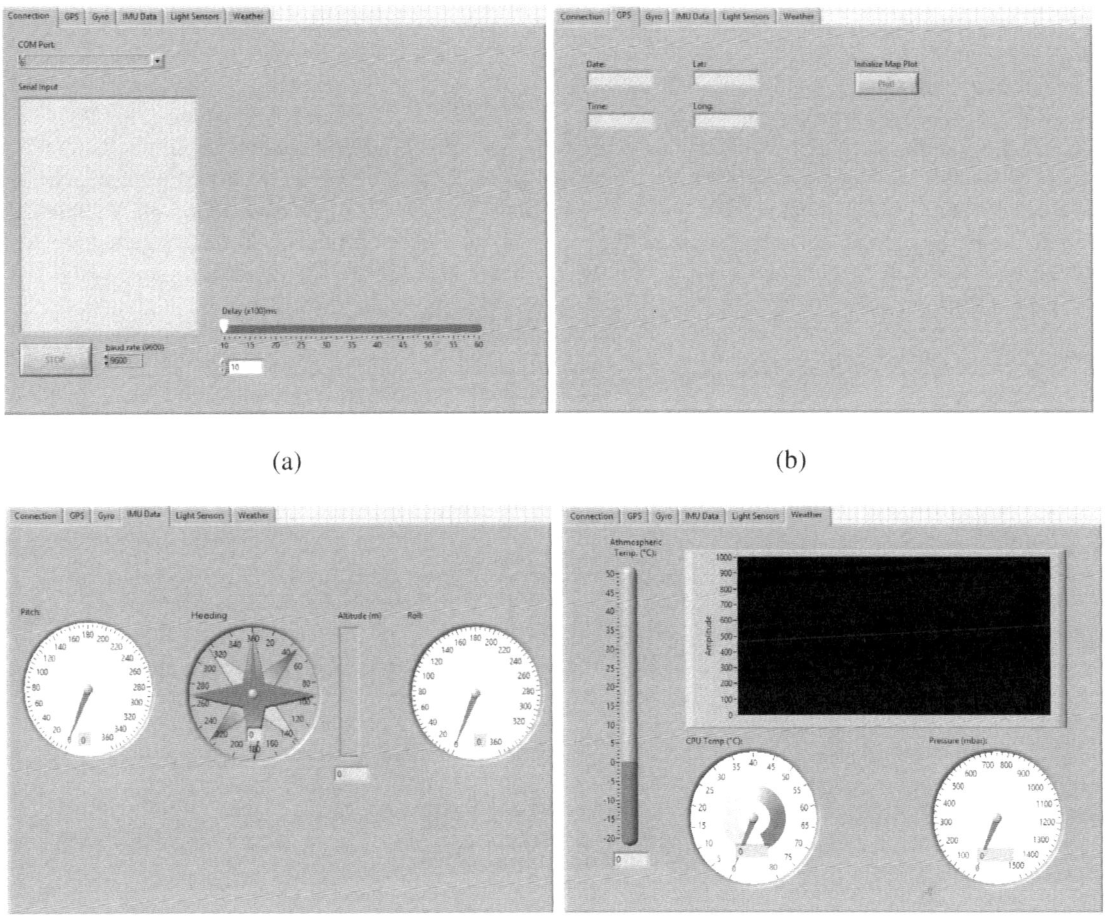

(a) (b)

(c) (d)

Figure 3(a-d): Different tabs of the Graphical user interface (GUI)

Figure 4 below summarizes the LabVIEW program used for the Ground Station. The serial connection opens with a predefined Baud Rate of 9600. The data that is being pulled from the XBee module connected to the computer is stored in a string where each data is separated using a comma delimiter. Then pattern match functions are used to parse each data before the commas. When the program stop button is clicked, the program is halted and the serial connection with the XBee module is closed.

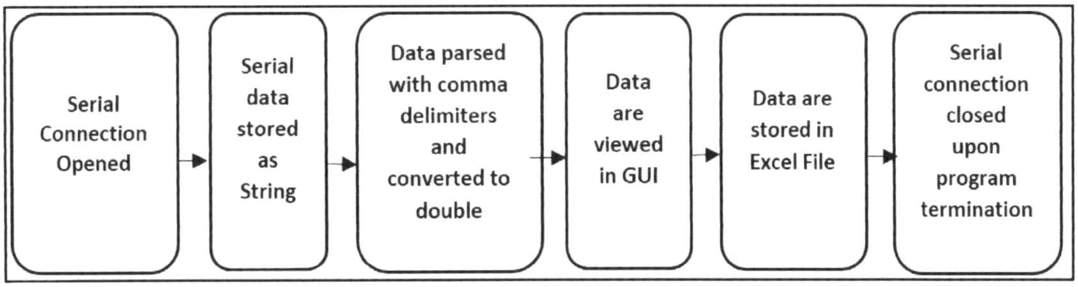

Figure 4: Flow-chart of the LabVIEW program

2.3. *Applications*

A cheap BalloonSat such as the one discussed in this paper can be used for several purposes. As it is easy to build and implement, multiple copies can be mass manufactured and flown in different areas of the country. Then collected data can be stored in a dedicated server for further weather pattern analysis. Students and researchers can benefit from these data to study the arrival of storms or other natural calamities. *Figure 5* below shows a plot of temperature against time, which was taken using the BalloonSat for ten hours.

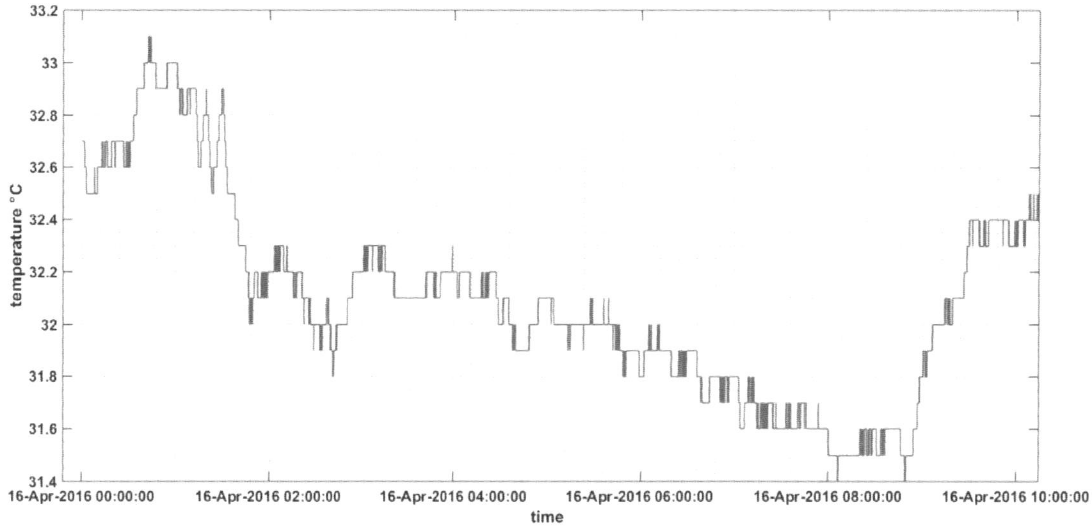

Figure 5: A plot of ten-hour time period temperature data

3. Image Analysis

3.1. *Image Thresholding and Boundary Detection*

Image thresholding is the simplest method of image segmentation. From a RGB image, thresholding can be used to create binary images [6] [7]. The simplest thresholding methods replace each pixel in an image with a black pixel (0) if the image intensity $I_{x,z}$ is less than some fixed constant T (that is, $I_{x,z} < T$), or a white pixel (1) if the image intensity is greater than the constant. Color images can also be thresholded. The approach is to designate a separate threshold for each of the RGB components of the image and combine them with an AND operation.

IOP Conf. Series: Earth and Environmental Science **38** (2016) 012009 doi:10.1088/1755-1315/38/1/012009

Figure 6: Color thresholding of an image of a park

Figure 6 above shows the original image as input and the green filtered image after thresholding. The image was originally taken by our BalloonSat. The pixels of the original image which have values within the defined range of RGB values are represented as white (1) and the rest as black (0). Later we used boundary detection to identify patches of greenery, which was inspired by the work of Maloof et al [8].

3.2. Vegetation Health Analysis

Image processing is very useful for agriculture; we can take an image of a vegetation field and match the leaf color with the leaf color charts to understand the Nitrogen fertilizer dosage of the field. C. Witt (2005) says that, the leaf color chart (LCC) as shown in *figure 7* helps the farmers to evaluate plant nitrogen (N) demand [9]. The LCC was made by calculating the leaf spectral reflectance measurement, this reflectance varies with the color of the leaf. The leaf color chart has four different green shades as shown in figure 7, so we took each shade and found its corresponding RGB composition in MATLAB. *Table 2* shows the RGB values obtained using MATLAB. An image taken from the BalloonSat can be processed using these values to find the nitrogen composition and identify plant types of a region saving farmers from manual inspection [10].

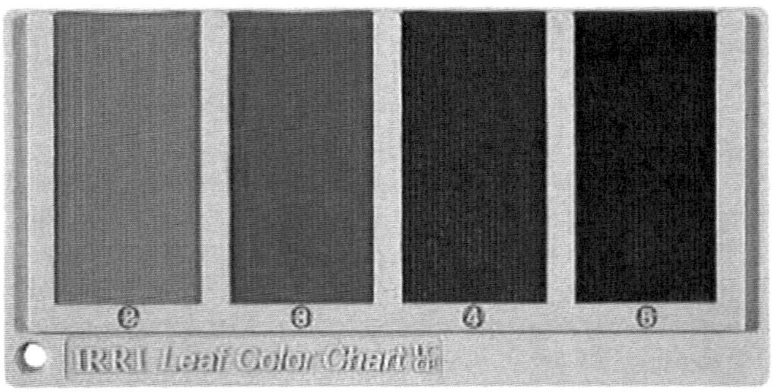

Figure 7: LEAF Color Chart (LLC), courtesy of Irrigated Rice Research Consortium

Table 2: RGB Spectrum of IRRI Leaf Color Chart (LCC)

Strip Number	Red	Green	Blue	Nitrogen Level
2	103-255	145-255	0-12	Critically low
3	83-98	145-165	0-30	Below average
4	0-106	111-113	0-30	Above average
5	0-104	0-105	0-62	Surplus amount

4. Conclusion

Space research is still a new field for a developing country. Before jumping to large-scale satellites for research purposes, Students can approach BalloonSat such as ours to collect data for analysis and implementation. This paper focuses on the acquisition of data imitating a real world satellite and implementing those data in our activities such as agriculture and weather. The acquired information could help farmers improve their yield of crop production; weather can also be predicted using a BalloonSat warning farmers and villagers of incoming natural disasters before it is late. BalloonSat could provide a developing country with the opportunity to approach space research, improve agriculture and weather prediction.

References

[1]. Garber, Steve. "Sputnik and The Dawn of the Space Age". History. NASA. Retrieved 6 May 2013.
[2]. K. Mason, "Rocket Lab And Balloonsat: The Importance Of Student Based Hands-On Experiments In the Classroom," 2009.
[3]. "Small satellite costs," Delft University of Technology, 2016. [Online]. Available: http://www.lr.tudelft.nl/en/organisation/departments/space-engineering/space-systems-engineering/expertise-areas/spacecraft-engineering/design-and-analysis/cost-engineering-cost-estimation/small-satellite-costs/. Accessed: May 17, 2016.
[4]. C. Koehler and C. S. G. Consortium, "DigitalCommons@USU - AIAA/USU conference on small satellites: BalloonSat: Missions to the edge of space," 2002. [Online]. Available: http://digitalcommons.usu.edu/smallsat/2002/all2002/55/. Accessed: May 17, 2016.
[5]. Keys, T. A., Jones, C. N., Scott, D. T. and Chuquin, D. (2016), A cost-effective image processing approach for analyzing the ecohydrology of river corridors. Limnol. Oceanogr. Methods. doi:10.1002/lom3.10095
[6]. Shapiro, Linda G. & Stockman, George C. (2002). "Computer Vision". Prentice Hall. ISBN 0-13-030796-3
[7]. J. Saghri et al., "Honors undergraduate research journal," 2009. [Online]. Available: http://digitalcommons.calpoly.edu/honors/vol2/iss1/6/. Accessed: May 17, 2016.
[8]. Maloof, M., Langley, P., Binford, T., Nevatia, R., and Sage, S. (to appear). Improved rooftop detection in aerial images with machine learning. Machine Learning. http://www.kluweronline.com/issn/0885-6125.
[9]. C. Witt, J.M.C.A. Pasuquin, R. Mutters, and R.J. Buresh. (2005) "New Leaf Color Chart for Effective Nitrogen Management in Rice"- Better Crops/Vol. 89 (2005, No. 1)
[10]. M. Miyamoto, K. Yoshino, T. Nagano, T. Ishida, and Y. Sato, "Use of balloon aerial photography for classification of Kushiro wetland vegetation, northeastern Japan," *Wetlands*, vol. 24, no. 3, pp.701-710, Sep. 2014.